The Coral Battleground, first published ~~~~~~~ ~ ~~~~ccessful struggles of Judith Wri~~~~~~~~~~~~~~~~~~~~~~ Barrier Reef from exploratory ~~~~~~~~~~~~~~~~~~~~ ;. Today, with UNESCO on the ~~~~~~~~~~~~~~~~~~~~ itage in danger', Wright's work re~~~~~~~~~~~~~~~~~~~~ irms in the new battles against pc~~~~~~~~~~~~~~~~~~~~

~~~~oy AM

Judith Wright's recount of the heroic early battles for the reef exemplifies the incredible achievements of a passionate few, who with vision and determination were able to succeed against the odds. This book is the stuff of legends and is a must read for all those who consider themselves environmental custodians. An extraordinary story in itself, *The Coral Battleground* now takes on an even more pertinent meaning as the Great Barrier Reef faces its biggest threat since the oil rigs of the 70s. May Judith's story inspire a new generation to fight for the reef!

–Bob Irwin, *Bob Irwin Wildlife & Conservation Foundation Inc.*

Political intrigue, shifting allegiances, dirty deeds, and more, Judith Wright records the environmental struggles of the 1970s, and the people's movement to preserve the Great Barrier Reef – for a time. Today, as the reef faces new threats, her book provides inspiration, and a how-to guide for a new generation of activists, for whom the beauty of the reef matters most deeply.

–Rosaleen Love, author of *Reefscape*

This edition of Judith Wright's lucid and compelling account of the fight to save 'the reef' comes with added value – a publisher's preface, a new foreword [by Margaret Thorsborne AO] and Judith Wright's own prophetic warning that such victories are never really won. Today's campaigners will find inspiration in Judith Wright's persistent vision in the face of great odds.

–Margaret Moorhouse, *Alliance to Save Hinchinbrook Inc.*

Wonderful and timely to see this lovely new edition of a classic book that every Australian should own. In it Judith tells the inspiring David and Goliath story of how she and a group of friends took on massive forces of greed and destruction that threatened the existence of the Great Barrier Reef and won. Australia urgently needs this new call to arms.

–Iain McCalman, author of *The Reef: A Passionate History*

Judith Wright, Arthur Fenton, John Büsst

Judith Wright is considered to be one of Australia's greatest poets; she was also an ardent conservationist and activist. Over a long and distinguished literary career, she published poetry, children's books, literary essays, biographies, histories and other works of non-fiction.

Her commitment to the Great Barrier Reef began in 1962, when she helped found the Wildlife Preservation Society of Queensland. She went on to become a member of the Committee of Enquiry into the National Estate and life member of the Australian Conservation Foundation.

Judith Wright worked tirelessly to promote land rights for Aboriginal people and to alert non-Aboriginal Australians about the awful legacy Europeans have left upon Aborigines and the land. She wrote *The Cry for the Dead* (1981), *We Call for a Treaty* (1985) and *Born of the Conquerors* (1991).

Judith Wright was awarded many honours for her writing, including the Grace Leven Award (twice), the New South Wales Premier's Prize, the Encyclopedia Britannica Prize for Literature, and the ASAN World Prize for Poetry. She received honorary degrees (D.Litt.) from the Universities of New England, Sydney Monash, Melbourne, Griffith and New South Wales and the Australian National University. In 1994 she received the Human Rights Commission Award for *Collected Poems*.

Other books by Judith Wright

Poetry:
Birds: Poems by Judith Wright (2003)
The Nature of Love (1997)
A Human Pattern: Selected Poems (1996)
The Generations of Men (1996)
Collected Poems (1994)
The Flame Tree (1993)
Phantom Dwelling (1985)
The Double Tree: Selected Poems 1942–1976 (1978)
Fourth Quarter (1976)
Alive: Poems 1971–1972 (1973)
The Other Half: Poems (1966)
Five Senses: Selected Poems (1963)
Birds: Poems (1962)
The Two Fires (1955)
The Gateway (1953)
Woman to Man (1949)
The Moving Image (1946)

Fiction/Short Stories:
The Nature of Love (1966)

Children's Fiction:
Tales of a Great Aunt (1998)
The River and the Road (1966)
Range the Mountains High (1962)
The Day the Mountains Played (1960)
Kings of the Dingoes (1958)

Non-fiction:
Half a Lifetime (1999)
The Coral Battleground (1977, 2nd edition 1996)
Going on Talking (1992)
Born of the Conquerors: Selected Essays (1991)
We Call for a Treaty (1985)
The Cry for the Dead (1981)
Conservation: Choice or Compulsion? (1975)
Because I was Invited (1975)
Conservation as an Emerging Concept (1970)
Henry Lawson (1967)
Preoccupations in Australian Poetry (1966)
Charles Harpur (1963)
The Generations of Men (1959)
William Baylebridge and the Modern Problem (1955)

JUDITH WRIGHT
THE CORAL
BATTLEGROUND

NEW EDITION

SPINIFEX

New edition published by Spinifex Press, 2014.
First published by Thomas Nelson (Australia), 1977.
Second edition published by Angus & Robertson, an imprint of Harper Collins, 1996.

Spinifex Press Pty Ltd
504 Queensberry Street
North Melbourne, Victoria 3051
Australia
women@spinifexpress.com.au
www.spinifexpress.com.au

Cover design: Deb Snibson
Typesetting: Palmer Higgs Pty Ltd
Printed by McPhersons Printing Group

National Library of Australia Cataloguing-in-Publication data:

Judith Wright, 1915–2000, author.
    The coral battleground/Judith Wright.
    3rd edition
    9781742199061 (paperback)
    9781742199016 (ebook: PDF)
    9781742199030 (ebook: epub)
    9781742199023 (ebook: Kindle)
    Includes bibliographical references and index.
    1. Natural resources – Queensland – Great Barrier Reef. 2. Environmental protection
    – Queensland – Great Barrier Reef – History. 3. Conservation of natural resources –
    Queensland – Great Barrier Reef. 4. Great Bather Reef (Qld.). I. Title.

333.709943

This project has been assisted by the Australian Government
through the Australia Council for the Arts, its arts funding
and advisory body.

To the memory of John and Taff
and to everyone who helped

Contents

Publishers' Preface

In 1975, the Great Barrier Reef Marine Park Authority was established. This was the result of many years of activism on the part of artists, poets, ecologists and students. They were called cranks, crackpots and worse, but today we can see the results of their activism in something very precious: the Great Barrier Reef Marine Park, the first such national park in the world.

In 1981, the Great Barrier Reef was selected for inclusion as a World Heritage Site. This means that it is a significant and unique place that should be protected for the future of all humanity. It is deemed, therefore, not a place to be turned into an industrial theme park, dotted with oil wells, dumped with polluting substances or overused in such a way as to threaten its existence. In 2006, it was included in the Australia's Biodiversity Action Plan.

What is so special about the Great Barrier Reef? Both of us live near the Reef and have seen its underwater beauty. For many people, perhaps a single dive or a holiday in which they snorkled the shallow coastline reefs, is sufficient to make them treasure the Reef. It is spectacularly beautiful. The colourful fish, the incredible shapes of corals, the intricate textures. Diving gives one a sense of freedom, of weightlessness, as close to flight as most humans will get. This is what Judith Wright meant when she said, 'the Great Barrier Reef is still the closest most people will come to Eden.' She describes what she saw wandering along shoreline pools on Lady Elliot Island:

I myself had seen only a very small part of it, in the fringing reef of Lady Elliott Island many years before the battle started. But when I thought of the Reef, it was symbolised for me in one image that still stays in my mind. On a still blue summer day, with the ultramarine sea scarcely splashing the edge of the fringing reef, I was bending over a single small pool among the corals. Above it, dozens of small clams spread their velvety lips, patterned in blues and fawns, violets, reds and chocolate browns, not one of them like another. In it, sea-anemones drifted long white tentacles above the clean sand, and peacock-blue fish, only inches long, darted in and out of coral branches of all shapes and colours. One blue sea-star lay on the sand floor. The water was so clear that every detail of the pool's crannies and their inhabitants was vivid, and every movement could be seen through its translucence. In the centre of the pool, as if on a stage, swayed a dancing creature of crimson and yellow, rippling all over like a windblown shawl.

That was the Spanish Dancer, known to scientists as one of the nudibranchs, a shell-less mollusc. But for me it became an inner image of the spirit of the Reef itself.

Judith Wright came to writing her book as a poet. She worked alongside the artist John Büsst, and ecologist Len Webb. They were early members of the Queensland Wildlife Preservation Society. Their activities were to lead them to court cases, local campaigning, getting the word out to the media and working with scientists from around the world. Their battle ended with a two-year long Royal Commission into the Great Barrier Reef because oil companies, hand-in-hand with the Queensland state government, were preparing to make test drillings in the Great Barrier Reef. The level of support they had was indicated by 35 barristers – including five QCs – who offered to work without fee during the Royal Commission.

Drilling for oil on the Reef seems unthinkable today because we have had nearly forty years of protection due almost totally to a small group of people who were determined not to let this happen. But we cannot rest on these laurels because the Reef is under threat again. This time it is approval by the Federal Minister for the Environment for the dumping inside the Great Barrier Reef Marine Park of around three million cubic metres of seabed dredged in order to build three shipping terminals as part of the prospective coalport at Gladstone on the Central Queensland coast.

In 2013, the ABC reported that two of the board members of the Great Barrier Reef Marine Park Authority (GBRMPA) had conflicts of interest because of links to resource companies. Joh Bjelke-Petersen and his Minister for Mines, Mr Camm had conflicts of interest in the 1970s which then and now are called 'irrelevant' by those who have a stake. Having pecuniary interest in resource companies, or companies that benefit a politician, does affect the judgement of those in positions of power.

Part of the problem is that prior to 5 September 2012, the authority had a strong policy line that it would 'not support port activities or developments in locations that have the potential to degrade inshore biodiversity'. In 2012, this was weakened considerably so that now it is only 'the *potential impacts* on inshore biodiversity [that] should be a key consideration.'[1] Such gouging of policy is happening in a wholesale way under the combined assault of the Campbell Newman state government and Tony Abbott's federal government. Between them, these governments are overhauling environmental legislation and anything that they insultingly deem 'green tape' is eliminated. But these so-called 'green tape' laws are what have kept some of our precious environments relatively free from pollution and other destructiveness.

While history is not repeating itself – because new strategies are used by decision makers – what is clear is that this is a continuous battle. Where there is no law to prevent overuse or dumping or drilling, then exploitation is pushed to the boundary until another ecosystem fails or a species is threatened. Then a new battle has to be fought. If governments respected the 'precautionary principle', activists and ordinary people would not always be on edge. The precautionary principle was written into the 1992 Rio Declaration on Environment and Development and signed by the participating Heads of States, including Australia.

The precautionary principle requires that where serious or irreversible damage is a threat, lack of scientific certainty should not be used to avoid measures that protect the environment.[2] Drilling test oil wells, dumping of seabed sludge, and the continuous release of nitrogen-laden water from Clive Palmer's Queensland Nickel or pesticide-laced agricultural run off from banana and cane farms are clear examples of not abiding by the precautionary principle. The proposal to double the agricultural

[1] Duffy, Conor. 'Great Barrier Reef board members Tony Mooney and Jon Grayson accused of conflict of interest over links to mining firms.' ABC, 7.30 Report, 30 October 2013. <http://www.abc.net.au/news/2013-10-29/reef-board-members-in-conflict-of-interest-claims/5052558>
[2] Hawthorne, Susan. *Wild Politics: Feminism, Globalisation and Bio/diversity*. Spinifex Press, Melbourne, 2002, p. 386.

output of land in the Wet Tropics put by the Abbott government will also contribute to increased pollution in the Reef.

Today, there is much greater understanding of the importance of biodiversity than there was in the 1970s. The interconnection between different parts of the Reef, including activities that occur outside the boundaries of the Marine Park are now much clearer, as is the interconnectedness of land use (farming, industrial growth, urbanisation and tourism), changes to the littoral coastal areas (especially the development of resorts and increased tourism, including ecotourism) and simply the number of visitors to an area, including residential developments and commercial tourist attractions.

Some governments have tried to ameliorate these impacts through offering environmental offsets, but offsets are always a negative. The precautionary principle, if adhered to, would suggest a considerable slowdown so that misdirected growth does not damage ecosystems or lead to increased extinction and threatened species.

The area around Ninney Rise, where John Büsst was based, is one of only two places in the Wet Tropics where the rainforest meets the reef. This rainforest is home to the southern cassowary, a large flightless bird who is in a symbiotic relationship with the rainforest. This is well summarised in the traditional saying of the Indigenous owners of the land, the Djiru:

no wabu, no wuju, no gunduy
no forest, no food, no cassowary

When rainforest seeds drop, the cassowary eats the seeds and disperses them as it moves through the rainforest. The rainforest in turn provides food and shelter for the cassowary.

When the rainforest is healthy, the reef is not suffering major run off, but rather both connect. Cassowaries also eat crustaceans on beaches at low tide and changes to the ecosystem will also affect them.

Similarly, mangroves in the intertidal zone – that space between land and sea – provide sites for fish nurseries. When dieback occurs – as at the mouth of the Johnstone River near Innisfail – these marine nurseries come under threat. The marine nurseries provide the next generation of fish who live in the Reef.

Climate Change is having an increasingly obvious impact on our environment. For example, Mission Beach has had two Category-5 cyclones in five years when the more usual previous pattern was a Category-5 cyclone roughly every twenty years. Global warming is accompanied by rising sea levels which in turn causes erosion of the foreshore. It also increases the temperature of the water. Seagrass meadows are being destroyed through increased activity throughout the Reef including more tankers passing through because of the expansion of the port at Gladstone. Seagrass meadows are essential habitat for dugongs and sea turtles, both endangered species. Increased freighter traffic combined with the dredging of the seabed and the dumping of the sludge inside the Marine Park could take the Reef to the brink.

While all this looks hopeless and there are days when despair at the postmodern, globalised and growth-centred economic policies make you throw up your hands, there are also reasons to be hopeful. Activists need to remain vigilant in order to resist proposals such as the Queensland government's latest idea to restrict who can object to mining proposals (limiting it to landowners in the immediate vicinity).

We do understand much more about biodiversity and inter-connectedness and we can talk to one another quickly over the internet and mobile phone services. There are new generations of activists with knowledge following in the footsteps of Wright, Büsst and Webb; there are artists and writers who care and whose work is inspired by nature; there are new generations who so far have taken the Great Barrier Reef for granted. With energy and action from concerned people, from those whose actions in the past helped – trades unions, ecologists, scientists and politicians – we could ensure that this extraordinary environment is retained and is there for future generations to wonder at.

In 2014, there are a number of hopeful actions taking place, such as:

- Brisbane-based Environmental Defenders Office, Queensland is taking legal action on behalf of the North Queensland Conservation Council with the support of protest group GetUp Australia to the administrative appeals tribunal.[3]
- The Environmental Defenders Office, North Queensland is calling for a change in Australian law, allowing for outstanding ecosystems to be granted legal personality. Legal personality would provide the necessary rights to protect the essence of that environmental entity. This would level the playing field between regions like the Great Barrier Reef World Heritage Area since inanimate constructs like corporations already have that status.[4]

Judith Wright in *The Coral Battleground* shows how a small group of dedicated activists can change history. Our hope is that it will be an inspiration for young people to join the fight to save the Great Barrier Reef, on that same coral battleground.

[3] <https://www.getup.org.au/campaigns/great-barrier-reef--3/reef-fighting-fund/reef-fighting-fund>
[4] <http://www.edonq.org.au/Campaign.html>

We are greatly indebted to Meredith McKinney and the estate of Judith Wright for granting us permission to republish *The Coral Battleground*. Thanks also to Suzanne Bellamy for her assistance. We are delighted with Margaret Thorsborne's Foreword to the new edition bringing her own personal contact with Judith Wright and others in the Wildlife Preservation Society of Queensland to the fore. Margaret knew both John Büsst and Judith Wright and, together with her husband Arthur, has been a tireless activist for the Reef not least through her activity as part of the Wildlife Preservation Society of Queensland. We also thank Suzie Smith, Secretary of the WPSQ, Cassowary Coast–Hinchinbrook branch, and Margaret Moorhouse whose own organisation, the Alliance to Save Hinchinbrook Inc., has kept developers away from the Reef and continues to contribute to its well being. Finally, we thank Liz Gallie for long conversations about matters concerning the Reef and Ninney Rise and for her commitment and passion to seeing the Great Barrier Reef and the Wet Tropics survive.

Susan Hawthorne and Renate Klein
Publishers, Spinifex Press
March 2014

Foreword, 2014
by Margaret Thorsborne, AO

By the early 1960s, there was a growing feeling of unease and helplessness at the unrestrained destruction of the natural world around us.

Four people who were keenly aware of this – poet Judith Wright, wildflower artist Kathleen McArthur, naturalist David Fleay, and publisher Brian Clouston – combined in 1962 to form the Wildlife Preservation Society of Queensland (WPSQ)[1] the first organisation in the state devoted to wildlife conservation.

The first test came when John Büsst, who had formed the Innisfail branch of the Society, saw a notice about a proposal to mine Ellison Reef for limestone. Simpler days back then in 1967: as Dr Don McMichael later told Suzie Smith of WPSQ, he had to speak very slowly as he gave pivotal evidence in the Innisfail Mining Warden's court, because the policeman recording the proceedings was writing carefully in longhand – proceedings which were truly momentous for the Great Barrier Reef's future: a precedent refusing an application to mine coral.

John Büsst was based at Ninney Rise, where he built a house at Bingil Bay in Far North Queensland near Mission Beach. From 1967 until his untimely death in 1971, John Büsst singlemindedly devoted his life to advocating for protection, not just for those coral reefs that might have been mined for lime, but for the entirety of the greatest barrier reef system in the world. This was, as Judith Wright later wrote in her

---

[1] <http://www.wildlife.org.au/>

Introduction to *The Coral Battleground*, 'the battle to save that thousand-mile stretch of incomparable beauty from the real destroyers – who are ourselves.'

The culmination of these campaign efforts led by Judith Wright was the declaration of the Great Barrier Reef Marine Park in 1975 in which there would be no drilling for oil. This was followed in 1981 by the Park's inclusion in a larger area for inscription on the list of world heritage properties, at the request of the Australian and Queensland governments. It was to be called the Great Barrier Reef World Heritage Area.

True to John Büsst's conception, it was not just 'the reef' as we had come to know it, but all that the coral reefs needed – the seagrass meadows, the mangroves, the coastal shoals, the marshes, the benthic communities, the seabirds and shorebirds, the dugongs and dolphins and whales, the turtles and a myriad species of fish – everything between the outer barrier reefs and the mainland shores: one great complex ecosystem and its integrity protected in perpetuity. We thought it a great victory that the Australian and Queensland governments had made solemn undertakings to protect the new world heritage area 'to the utmost' of their capacities, according to the terms of the international World Heritage Convention.

Nevertheless, Judith Wright continued to reflect on 'the real destroyers – who are ourselves'. In her letter to me written just a few months before she died on 26 June 2000, she wrote:

> It is a long time since the Reef was won, and it still surprises me that ... it was won at all – if one can say that the present situation is a victory.

By then, Judith had become aware of many more emerging threats to the Great Barrier Reef World Heritage Area and the increasing signs of serious trouble – vanishing seagrass and shrinking populations of dependent species such as dugongs and turtles; infestations of crown of thorns starfish and episodes of coral bleaching. Since that time, we have all become aware of the overwhelming facts of human-induced climate change: increased seawater temperature, increased sea level, increased weather variability, and increasing ocean acidification – which corals cannot survive. Seagrass continues to vanish; dugongs no longer inhabit the Southern Great Barrier Reef waters as a viable population.

The beauty and wonder of the largest and most extraordinary coral reef on earth, so large its patterns can be seen from space with the naked eye, is now under such intense threat that the United Nations

Educational and Scientific Organisation (UNESCO) has warned the Australian and Queensland Governments that if they do not make a real improvement to the deteriorating condition of the Great Barrier Reef World Heritage Area, it may be placed on the list of 'world heritage in danger'.

As reported to the UNESCO by the World Heritage Committee (Mission Report June 2012): 'development pressures, reduction in water quality, and climate change are clearly impacting on the values of the property' and 'are expected to pose the greatest threat to the long term conservation of the property.' As the Committee expressed it, 'building resilience – through reduction of and/or elimination of other pressures – is crucial to ensure habitat areas are capable to adapt to a changing climate without going extinct.' In recent times the threats of catchment run-off, coastal development and ports have burgeoned, 'more than 65% of all coastal development proposals were made in the last five years,' that is, since 2007.

It is the Queensland mining boom which has driven the demand for bigger port facilities, more dredging, more seadumping of dredge spoil, more reclamation - all to serve more and bigger ships for which ever deeper channels are cut to reach navigable waters within the shallows of the Great Barrier Reef lagoon.

Perhaps the only threat not currently applicable to the Great Barrier Reef is oil exploitation, which, as the Mission Report noted '*is legally prohibited*' – true to the unprecedented vision, determined hard work, and extraordinary persistence of John Büsst and Judith Wright. To celebrate the 45th anniversary of the winning of that first, smaller but crucial success of Ellison Reef, ceremonies were held in 2012 at Ninney Rise. Eddie Hegerl of the Queensland Littoral Society (now the Australian Marine Conservation Society), who had taken part in the original campaign, dived Ellison Reef once more, finding it still healthy and recognising corals from his dives there in 1967.

In the latter part of 2000, I joined others on Mount Tamborine in the national park near Judith's old home 'Calanthe', named for the beautiful white flying dove orchid, to give homage to her memory, joining others in speaking about her life and the generosity of her friendship to so many people. The whole world owes so much to this great Australian poet and leader of the conservation movement. It was her passion for the natural world, brought to the easy understanding of people through her great gift and skill with words, which inspired and empowered the campaign which should have ensured the life and the beauty of the Great Barrier Reef for ever.

This re-issue of Judith Wright's book is timely. Not only because this great story will reach and inspire another generation of young Australians, but also because Judith's words will remind us all that although a battle may be won, it is never a victory while the real destroyers continue to wage war on the natural world that sustains us all.

Margaret Thorsborne AO is a conservationist and environmental activist. She is particularly known for her efforts, with her husband Arthur Thorsborne, in initiating the long-term monitoring and protection of the Torresian Imperial-pigeon, a migratory species which nests on islands near Hinchinbrook Island, Far North Queensland. More recently she has been involved in the struggle to protect Queensland's Wet Tropics World Heritage Area and species such as the Southern Cassowary, Mahogany Glider and Dugong.

Foreword, 1996
by Judith Wright

There are not many success stories in the attempts we make to save especially important elements of the natural world from our own greeds and needs. Here at the end of .the twentieth century, we have lost or destroyed a great deal already, and we know that much more is likely to vanish. But the story of the rescue of the Great Barrier Reef still throws a light on the present and gives hope for the future, and because of the rescue many people have been able to experience and enjoy the marvellous stretch of sea and reefs and islands, and the intricate patterns of living beings, which make up its existence.

This book was written at a time when it still seemed uncertain that the rescue really would take place. The ugly demands of industry, employment and trade seemed beyond control. When Prime Minister Whitlam acted to declare the setting up of the Great Barrier Reef Marine Park and its Authority; under the establishing Act of 1975, with the goal of 'providing for the protection, wise use, understanding and enjoyment of the Great Barrier Reef in perpetuity through the care and development of the Great Barrier Reef Marine Park', political and industrial opposition to its existence was intense. It seemed laughably unlikely that successive federal, let alone state governments would not undo the Act. The values of the Reef in oil, minerals and other dollar equivalents might be questionable and unproved, but the forces which wanted to exploit them were immense.

The pressures on its future remain heavy, but the counter-pressures have proved stronger. One of these is the determination and devotion of many people who have seen the Reef and love it, not for the dollars that might be squeezed from it in production and tourism but for its overwhelming and marvellous existence. Though its brilliant waters have been dulled and darkened here and there by unwise and greedy uses and human and industrial forms of pollution, the Great Barrier Reef is still the closest most people will come to Eden. That it has survived the various industrial threats of the past quarter century is due not only to the existence of the Act of 1975, but is also because such people have decided it should stay so, and because the governments, committees and council that make up the Great Barrier Reef Marine Park Authority, have agreed with them. That's a miracle in itself.

Moreover, in the most recent years of the Authority's existence, it has widened its scope in obedience to the change in the law that has at last recognised the prior claims of the original inhabitants and users of the Reef and of the Torres Strait, and added their groups to its administrative and management components. This will increase its influence and the respect in which it is held, and strengthen its decision-making and advisory elements.

To me, it's a kind of miracle that things have gone so well for the Great Barrier Reef. But I know that its survival is owing to a great deal more than luck and circumstance. Luck there has been – no big tanker has crashed in its passages, no plot by destructive forces has succeeded in breaking down its legislative and managerial defences, it has been very fortunate in the official appointments and scientific tasks that support it. If disasters in the shape of weather, accident and climate change lie ahead, the work done already has shown what can be done to shield it from such dangers and has proved that people will agree, in the event, to supplying the help it needs.

The battle for the Reef will never be quite over, but in terms of the rewards already gained it has been, and is, more than worth while. That splendid stretch of the northeastern coast of Australia is an enrichment of human experience of the beauty of the world that is without parallel. The battle for the Reef stands to the credit not just of Australians but the human race.

Judith Wright
Braidwood NSW 1996

Foreword, 1977
by Judith Wright

I have chosen to tell the story of the Great Barrier Reef from the point of view of those actually involved in the battle to prevent the Reef from oil-drilling and limestone mining. Obviously, I have not had access to a number of sources which could have presented the story from the other side – e.g., the records of the Queensland Mines Department.

This book does not pretend to be an exhaustive survey based on all the literature available on the Great Barrier Reef and its problems, nor even a complete account of all the events leading up to the present situation. To have attempted to give an account even of the evidence given to the Royal Commission into Petroleum Drilling in Great Barrier Reef waters, would have overloaded it.

The book is thus not a reference work, and I have not attempted to give full references for the material used – especially for the newspaper reports and articles which covered the period. However, I wish to acknowledge the role of the press, both in reporting and in producing special articles and surveys, throughout the whole campaign from its inception in the Ellison Reef case. I make particular acknowledgement to *The Australian*, whose full and faithful coverage was a most important factor in the amount of public interest and information on the whole question.

Introduction

This story has no real beginning and no one knows what its end will be. It is a part of the history of the Great Barrier Reef, that great complex structure of coral reefs and living organisms that stretches 1,200 miles along the coastline of Queensland, sometimes near the shore, sometimes many miles out to sea, in a scattered or concentrated succession all the way from the northern tip of Fraser Island to beyond Cape York in the north. The Reef's story goes back far beyond the time when Cook's *Endeavour* found its way northwards to Tones Strait and the first white men stared down in awe at its crags and underwater gardens, and navigated among them as reef after reef threatened their keels.

> Flowers turned to stone! Not all the botany
> Of Joseph Banks, hung pensive in a porthole,
> Could find the Latin for this loveliness,
> Could put the Barrier Reef in a glass box
> Tagged by the horrid Gorgon squint
> Of horticulture. Stone turned to flowers
> It seemed – you'd snap a crystal twig,
> One petal even of the water-garden,
> And have it dying like a cherry-bough.
>
> 'Five Visions of Captain Cook',
Kenneth Slessor

Ever since that voyage of the *Endeavour*, ships have passed through the Reef's waters between the coastline and the Outer Barrier, and many of them have been lost. From the wooden sailing-ships to today's great steamships, cargo vessels, and now the huge tankers, they fear the Reef and its treacherous waters and weather. But if the Great Barrier Reef could think, it would fear us more. We have its fate in our hands, and slowly but surely as the years go on, we are destroying those great 'water-gardens', lovely indeed as cherry-boughs in flower under their once-clear sea, but far more complex, far more alive, teeming with myriads of varied animal lives. The Great Barrier Reef has killed many ships and men over the years; but it has drawn and fascinated everyone who has ever seen it, and men have fallen in love with it as well as despoiled it. This is the story of the battle of a few people who loved it; the battle to save that thousand-mile stretch of incomparable beauty from the real destroyers – who are ourselves.

The story will say little of the Reef itself. It is a political story, but also it is a story of people in their interactions. Its complications stretch far beyond Australia, for the Reef has its enemies, and its lovers, in many countries and in the whole complicated structure of the world, its industry, its science and its commerce.

For many thousands of years it saw no men at all; the Aborigines were not seafarers and few if any of the Pacific Island voyagers can have been blown so far as to see it. But the *Endeavour's* voyage brought it into the world and exposed it to all the dangers of a civilisation that lives by exploiting everything in land and sea. The Reef will never be alone again until that world ends; and then it may not be a living assemblage. It may be a dead, blackened and crumbling hedge of limestone rock, its gardens withered and its creatures decimated.

Meanwhile, there is a story to be told.

1

The First Manoeuvres

My qualification for writing this book is that I was privileged to be one of the people who fought the battle for the Reef itself. My own contacts with it have been few. I was ten years old when I passed through the Reef waters, in a steamer sailing north, and I did not see the water-gardens. I remember the marvellous blue of calm waters, the green islands of the Whitsunday Passage rising out of them and passing by, and not much more.

I did not see it again until in 1949 I spent a few weeks on Lady Elliott Island, the southmost of the Reef's coral isles, staying in a lighthouse cottage. The island was already spoiled; its guano had been stripped and shipped away, and little was left of the vegetation except a few pisonia trees nibbled by the herd of wild goats left there for the lighthouse meat supply long years before. But the offshore reef was still beautiful, and I wandered over it amazed at the colours of the corals, the shellfish and the tiny darting fish and crimson and blue slugs and stars and clams in its pool-gardens, and stared down from a small boat at its shelfs and coral crags. I fell in love with the Reef then, through that small and southmost part of it.

Fourteen years later, I helped to form a conservation society in Brisbane, the Wildlife Preservation Society of Queensland, and here my part in the Reef's story begins, and my involvement with many others who knew and loved the Reef, or who wanted something from it.

Our society was a small one then; it started with less than a hundred members, and our chief aim at that time was to start a magazine which would be a forum for conservation and would educate people in the value of wildlife. It was a hard struggle to get this magazine going, and it took most of our time. It was an education for us too. We met and heard from many people and began to learn a great deal about what was happening to Australia, and particularly Queensland, as the rush of 'progress' and 'development' increased. We learned to dislike the sound of those two words.

Most of us in the society at that time were people who were concerned and troubled at the destructiveness of much that was happening, but had no professional qualifications in biology. However, we had to teach ourselves a great deal, and ask questions, and before long we were joined by people who did know a good deal: botanists, biologists, naturalists. Letters began flowing in, and we reached out to other people at a distance; we read and we discussed. We had one lucky hit early on, when we decided that we needed an ecologist to help us get our priorities right; this came to the notice of one of the few qualified ecologists in Australia, Dr Len Webb. He was then working in the rainforests of Queensland's far north, and when he came to Brisbane and offered his help, we seized on him and made him a vice-president at once.

In northern Queensland he had met many people who were interested in the fate of the northern rainforests, and one of these, John Büsst, was also a lover of the Great Barrier Reef. John's involvement with the society came later, but he was to be a chief figure in the story.

In 1963, we became concerned over reports of what was happening to the Great Barrier Reef. Coral collectors, shell collectors and tourist interference were increasing rapidly, and some people in the north were unhappy with this. The photographer Noel Monkman, who with his wife Kitty was living on beautiful Green Island offshore from Cairns, was one of our correspondents. He was troubled over the future of his beloved reefs and islands. We began to wonder whether anything could be done to help, and the idea of the Reef's becoming a great underwater park was brought up.

It was an idealistic notion at the time; and we were duly pooh-poohed by most of the people we approached with the suggestion. But we learned something from that early approach, nevertheless.

It seemed that the responsibility of looking after the Reef, if there was such a responsibility, lay with the Queensland State Government, and that the Commonwealth Government was not particularly interested in what happened to it. The Queensland government had made a few regulations on Reef exploitation, notably about the removal of live coral, and had declared some protection. Wistari Reef, off Heron Island, was the main protected area, and there was a small scientific research station on Heron Island itself, funded by meagre grants, and run by a body called the Great Barrier Reef Committee, a small, non-profit-making scientific body, for many years the only independent organisation specifically concerned with the Reef, and at this time the only authority on matters of Reef administration and biology.[1] Beyond that area, and Green Island, information was scanty about what controls there were even over the taking of corals.

We made a few approaches on the question of marine national parks – of which at that time there were few in the world. The departments concerned were not interested; the zoologists and marine biologists we spoke to were of the opinion that nothing needed to be done. They told us that shell-collecting and coral-taking were negligible and that the Reef resort proprietors looked after their own areas and would not welcome our notions or our interference. We were too busy to pursue the idea, but we put it away in the hope that we might get more encouragement later, and then we might think once again about the problems our northern friends told us were real enough.

The next step came in 1966, a year when, from being a small collection of cranks labelled as anti-progressive visionaries, conservationists (as we are now called) began gathering strength. The society already had several branches in areas outside Brisbane and was moving into further fields: we now had more than 600 members and more people were willing to take an active hand. It was easy to see that the shibboleths of growth and progress needed a balancing force, if the future was going to be lived in a world fit for humans.

Queensland, always a state that longed for more of growth and progress, was forcing its pace. It had few areas of reserved land where its great variety of plants and animals could look forward to protection, and it was wide open to every proposal for development, mining, industry and settlement. The northern rainforests were fast being felled, often with disastrous results in erosion, silting of streams and estuaries, and poverty-stricken farms. Len Webb, a vital and urgent man with a love for the magnificent forests he studied, travelled to and fro, talking to people and making himself unpopular, but also being heard by those with foresight.

He would come back to Brisbane imbued with the tragedy of the

---

[1] See Chapter 2, pp 19–20 *passim.*

forests and keep us all alerted to their needs. He and his small team had studied many of the forest areas and made plans and proposals for national parks, few of which got a hearing from the government. The rainforest continued to be felled and burned, and plants and animals unknown, or almost unknown, to science, and never to be replaced, went up in smoke. Progress was the cry, and progress we got, no matter how destructive and planless.

At Innisfail Len Webb had an ally in John Büsst. I met John and we talked. He was an artist, Melbourne-born, who had moved to one of the lovely islands offshore from Innisfail, Bedarra, and with his wife Alison had lived there many years, painting and boating and swimming and learning about the forests and reefs. Now he had moved back to the mainland at Bingil Bay and built a house there; fired by Len on their expeditions through the rainforest he had formed an organisation consisting only of himself, as president, secretary and treasurer, with its own letterhead, to help save some of the Innisfail forest.

He was a slender, enthusiastic man full of laughter, a compulsive smoker and a lover of good company, a friend of the then Prime Minister, Harold Holt, who himself had a 'hideaway' house not far from El Arish. John had spent much time with the Holts on their northern holidays and they had talked of the Great Barrier Reef and its future; Holt was a convert to John's ideas on the need for protecting the Reef, and this was to be a beginning of a change in thinking in Canberra on the Reef, which had been left so long to itself and its distances.

There were others round Innisfail with an interest in the rainforest and the Reef areas and a concern for both, and now John decided to push for a branch of the Wildlife Preservation Society there. In August 1966 he called a meeting, and the branch came into existence with John as president, and Mrs Billie Gill, an amateur ornithologist of great vigour and enthusiasm, as secretary. Billie was another of the friends of Len's rainforest team, working with them often and identifying rainforest birds. A farmer's wife, she had a busy life of her own to attend to, but the birds were and are her first love, and she is well known to many ornithologists who have visited North Queensland and indeed to many who have not. I didn't meet Billie for another year or two, but when I did I found her a rare person, living up to and beyond Len's description. The Innisfail branch was off to a good start.

There were other moves in the south that were to be no less important. In 1965, the University of New England had held a seminar on Wildlife Conservation, and a number of us from Queensland had gone down to deliver papers, including Len and his offsider botanist Geoff Tracey, and myself. There we had talked with Dr Francis Ratcliffe of CSIRO in Canberra, who was working towards the establishment of a national conservation body. Len and many others were enthusiastic about the idea, and finally in late 1966 the Australian Conservation Foundation held its first meeting. Len and I were both invited to be members of the Provisional Council soon set up.

All of us began to feel we were no longer lone operators. We had now met many other people working in our field, we were full of the euphoria that comes to small embattled groups when the idea they are working for begins to break through; in spite of some internal doubts and disagreements, the conservation movement began to feel itself a happy few, a band of brothers and sisters, but with achievements ahead.

Also we were meeting more and more people from overseas, biologists and naturalists and conservationists, who had come to see Queensland's rich forests, reefs and animals and to warn us of the need for working to get protection for them and to tell us sad exemplary tales of what had happened overseas. We scarcely needed telling by now; but they were a further inspiration and very useful publicity. Many of them became our friends and correspondents.

In 1967, another new society had appeared at the University of Queensland, the Littoral Society of Queensland. Most of its members were young: students interested in marine and freshwater and like problems of biology, and more clued-up on problems of water pollution than we were then. They took up the question of marine national parks, and we got in touch with them at once and suggested cooperating.

We had had problems in the fragmentation of conservation groups; we were considered rather hotheads by elder groups whose main interests were in naturalist study or whose policies were already fixed. We were demanding action, and we were thought to be publicity-seekers; as for the magazine, few outside the society thought it had a chance of success, and indeed its early struggles had often discouraged us. But during 1966-67 we had managed to organise it onto a more secure footing; it was now edited by a good naturalist with a flair for publicity, Vincent Serventy, and it was going all over Australia, into schools and libraries, and acting as a forum everywhere, as we had hoped. We had time to turn some of our attention to wider problems.

Some of the young men from the Littoral Society came onto our Council; including Des Connell, working towards a higher degree in chemistry and especially interested in water pollution, and Eddie Hegerl, a collector of marine organisms in the Department of Zoology. Their knowledge and youthful enthusiasm set us off again on new questions. It was not long before we were to put all this new human equipment to its first big test.

In mid-1967 the Innisfail branch got in touch with us urgently. In our June newsletter, Des Connell had contributed an article on marine conservation, with special reference to Queensland's responsibility for the Great Barrier Reef.

'We in Queensland have particular responsibility,' he wrote, 'as we virtually hold the Great Barrier Reef in trust for future generations throughout the world.' He suggested that the declaration of marine national parks in particularly rich areas of the Reef was the most hopeful method of conservation; that this would not only provide new attractions for Reef visitors but areas where base stocks of fish could be protected for replenishing marine fisheries. And he pointed out that pollution was a real danger to the Reef. Huge fish kills from chemical pollution were being reported overseas; but 'the diversity of forms pollution can assume ... can vary from sediment stirred up in mining operations to radioactive fallout'.

That reference had caught John Büsst's eye at the right moment. He had discovered a limestone-mining application for removal of coral from Ellison Reef, offshore from Innisfail, advertised in the local papers. The Innisfail branch had lodged an objection. Would we do so, too?

We did; so did the Littoral Society. John approached the Australian Conservation Foundation and many others. The ACF too lodged a written objection. The fight was on. It was the first stroke in a battle which was to occupy our minds and time for years ahead. We were plunged at once into a mystifying controversy.

The grounds that John had chosen for objection were twofold: first, the danger to Ellison Reef itself, but further, the danger of establishing a legal precedent for mining that could lead to widespread commercial exploitation of the whole of the Reef.

Ellison Reef is an isolated outlier from the main body of the great barrier and lies inside it, eighteen miles east of Mourilyan Harbour. It was therefore closer to the mainland than the outer barrier, and fairly easy of access, at least in fine weather. It lies wholly underwater, but its upper surface comes close to tidal levels during low tide, and removing coral would not have been a difficult job, in calm weather. Coral is composed of almost pure carbonate of lime; where mainland lime must be mined and purified, coral limestone needs little treatment beyond crushing to be suitable for use in farming. The coast of northern Queensland, where conditions are suitable, is one of the most important sugar-growing areas. The cane-growers need a good deal of lime, and this can be expensive. An enterprising cane-grower, one D. F. Forbes, had seen an opportunity to get lime, as he claimed, at less expense to cane-growers, and had applied to dredge Ellison Reef for the purpose.

This was the first application of its kind relating to coral mining in Reef areas, and as such it had an importance far beyond its immediate aspect. Cane farmers all along the Queensland coast are an important group with political influence. If this application were to succeed, we could foresee many more such applications and a running series of battles all along the coastline.

Any form of mining or dredging, on such reefs, would create just the problem that the Littoral Society had pointed to, of siltation and water pollution, not only in the immediate area but along the currents that make an intricate network of patterns throughout the Reef, as far as the silt might be carried. Silting had been proven to damage living organisms, particularly sedentary ones such as coral. The Littoral Society had been working on an underwater survey in the Tweed River, where dredging had recently been taking place, and had seen what silting had done to the underwater ecosystem.

Some of the young biologists had also been working from the Heron Island research station, where various operations intended to make a boat-harbour for tourists had silted the offshore reef with consequent damage. They were deeply concerned at the possible implications of granting limestone-mining applications along the Reef, and the Littoral Society was alerted at once. Their cherished project of marine national parks might come to little if the Reef was subject to the damage caused by dredging and blasting for limestone.

We knew that in other mining situations, the establishment of a precedent for granting mining was important for subsequent applications. It seemed to us a clear case; if the Reef was to be protected, the Ellison Reef application was a spearhead for establishing the evidence necessary to show that limestone mining would harm it.

The Great Barrier Reef Committee's scientists, who ran the little Heron Island research station, were mostly based at the University of Queensland. John Büsst felt that the university's interest in marine research and in the station itself should make it a principal objector to exploitation which might damage the Reef as a whole. It was rumoured by this time that the Reef was endangered not only by limestone mining, but by numerous applications for oil prospecting – already the Swain Reefs, not far from Heron Island, were being drilled under a prospecting permit granted by the Queensland government.

We knew that the Great Barrier Reef Committee had expressed some concern about this; but applications for oil-drilling offshore were not advertised for objection, as limestone-mining applications were, and we felt that the Ellison Reef application offered a perfect opportunity for scientists to testify on the dangers of pollution to the Reef generally, and the vulnerability of the Reef to those dangers.

Accordingly, John wrote to the university, among many other institutions and societies which might be thought to have a special interest, and asked whether it would be among those who were opposing the application. By this time, he was employing at his own expense a firm of solicitors in Innisfail who were preparing the case to be heard at the end of September. This firm, Arnell & Arnell, entered into a good deal of correspondence with the objectors, on behalf of the Innisfail branch.

The university replied that 'it appears that the portion of the reef known as Ellison Reef … is dead and in consequence exploitation would not endanger living coral. In view of this, the University would not oppose the granting of the lease. Were the application to be made for mining of living coral, the matter would receive special consideration.'

Ellison Reef dead? How had it died, and when? We were all amazed, even uncertain. The question of pollution from dredging and blasting remained, whatever the condition of Ellison Reef, and we were not going to pull out of the case on such a ground, but we had to find out more about Ellison itself.

John Büsst took counsel with local fishermen and flew out over Ellison Reef. It looked underwater like any other reef, and the fishermen thought so too. So did Dr Barnes, a medical man with a special interest in the Reef through his work on the origin of the disease known as ciguatera. He was one man whom John had enlisted as a witness, since he knew a good deal of the Reef itself, and his work pointed to a possible cause of ciguatera, in eating carnivore fish which had been in areas where algal growths were concentrated on dead coral boulders. This disease was increasing among Dr Barnes' patients, as indeed all over the Pacific. But Dr Barnes was emphatic that Ellison Reef was no more dead than any other such reef.

Indeed, all coral reefs contain within their growth cycle much 'dead coral', broken by weather into coral rubble, lying on the reef surfaces as broken coral branches and sand. Such 'dead coral' is an integral part of Reef life and forms breeding areas not only for algae and small marine organisms, but for fish themselves. Like the dead leaves under trees in the forest, it has an important part to play in replenishing Reef life; it is not an expendable 'dead' product, but teems with life of many kinds. The University's reply bewildered us. But since the application was to dredge 'dead coral', it was obviously important to establish this, and also to establish that the whole of Ellison Reef was not composed of such dead areas. If the University and the Great Barrier Reef Committee were not disposed to give evidence to this effect, we had to establish it for ourselves. And it seemed, indeed, that they were not. All those we contacted refused to give evidence in the Ellison Reef case. Why?

John and his solicitors had by now collected a good number of written objections, but they had no witnesses to appear in the case except John himself, and Dr Barnes, neither of whom could be considered 'expert' in marine biology. The case was being heard in Innisfail, many hundreds of miles north of Brisbane; the marine biologists' refusal to appear meant that no 'expert witnesses' could be called on to support the objection. The end of September came, and the case came on.

It was now revealed by the Mining Warden that he would not accept written objections unless witnesses appeared for cross-examination. All the objections lodged by ourselves, the Littoral Society, and the Australian Conservation Foundation, would thus go for nothing. John's solicitors demanded a postponement of the case to get witnesses who would support the written objections, and this was granted.

It was now that we really began to appreciate the calibre of our Innisfail branch, and particularly of its president. John telephoned me in Brisbane with the news of the postponement. We had a few weeks to organise appearances and further evidence. John had given up hope of any marine biologist in Queensland; but we had another card or two.

We had an urgent meeting of our council, and the Littoral Society representatives made a suggestion. They were young men with no money to speak of, but they were experienced divers and some of them already had qualifications as biologists and marine collectors. If we could raise the money, the transport, the accommodation and some of the equipment, they were willing to apply for leave from the University to go north and organise a dive on Ellison Reef and a count of species.

We looked into our financial resources, they were slender enough but we felt we could do something. I rang John Büsst and asked if transport and accommodation could be arranged somehow. Yes, the divers could stay at Bingil Bay; he himself would fly south and talk to airlines and other organisations, to try to get free transport. He was optimistic that Sir Reginald Ansett, who owned Reef island resorts and had much interest in tourism to the Reef, would supply free air fares at least; and the manager of Avis Rent-a-Car, who was a friend, would supply the trucks and ground transport needed for the equipment and the divers.

Meanwhile, I telephoned Dr Don McMichael, the newly appointed Director of the Australian Conservation Foundation. McMichael had been in Queensland just before this, when the Foundation had organised a public symposium 'Caring for Queensland'. He was himself a marine biologist, and his address had dealt with the very question of marine national parks and the Great Barrier Reef. I asked him about Ellison Reef. Was it, in fact, a 'dead' reef?

Here we had a most unexpected stroke of luck. Most of Ellison Reef was inaccessible except to a diver; and it had been impossible to find any marine biologist who had worked on Ellison – or so we thought. We had tried hard enough … but Don McMichael, over the telephone, sounded surprised. Yes, he had worked on Ellison Reef. No, it was, as far as he remembered, quite a normal reef at that time – it was some years ago, but he knew of nothing that would have killed it, and indeed dead reefs, as such, were rare indeed. He had done some collecting there, and would look up his records.

Could he possibly come up to Innisfail for the hearing on 21 November and support the ACF's written objection, as an expert witness?

As he said later, at first he was doubtful. Ellison was only one of many similar reefs; there were thousands of such reefs all the way down from the Gulf of Papua to waters offshore of Rockhampton. But he promised to consider it.

At the ACF Symposium, he had argued that the Great Barrier Reef was so important, scientifically and otherwise, that no possibly damaging action should be taken in its waters until a thorough survey had been made and all the implications of action had been assessed. He had suggested that a major scientific survey should be begun without any further delay; and he had talked to marine biologists from the Fisheries Section of the State Department of Harbours and Marine, as well as to officers of the Great Barrier Reef Committee. All of them had been interested in the idea of a major survey, or had said they were. There was already a possibility that the Australian Academy of Science might be interested in helping to mount one, and that the ACF itself might sponsor it jointly with the Great Bather Reef Committee. McMichael applied to the Foundation for leave of absence to appear in the case. He suggested that the Foundation's objections should take two grounds: the scientific interest of Ellison Reef itself, and the need for a properly based survey of the whole of the Reef, in which the Foundation itself might be a moving sponsor.

The ACF, perhaps with some reservations in the matter of committing itself to becoming a sponsor for a survey likely to be far beyond its prospective means and organisation, granted Don his leave. He had discovered from his records that in his Ellison Reef research he had collected specimens of an especially rare shellfish, known from few other places, and this would be ammunition for his scientific evidence. He, too, wanted a chance to dive again on Ellison before presenting evidence.

In Brisbane, we were feverishly engaged in getting equipment and organising the Littoral Society's diving expedition. The university was not very co-operative about granting leave of absence, but it was being taken anyway. John telephoned from Melbourne and then from Sydney. He had run up against a blank wall with Ansett, but he had convinced the second airline, TAA, and they would provide air transport to Cairns. Avis would meet the plane at the airport and bring the equipment and the divers to Innisfail; Kodak Ltd promised to supply free film for the cameras. Somehow we scratched up enough equipment and money to back the expedition. Telephone lines ran hot between Brisbane and Bingil Bay, as Alison Büsst, who had been left to organise the accommodation and boats, told us how things were going. A friend of the Büssts, a local boat-owner named Perry Harvey, was organising transport out to Ellison.

Don McMichael was to be brought from Melbourne by TAA, free, and Avis would provide his transport; the cost to the ACF would be minimal.

The scuba expedition set off, and now we could only wait and hope for fine weather at Innisfail. November might easily bring rough rainy days with no chance for diving. The weather was just good enough, during those few days, to allow about thirty man-hours for the survey. Eddie Hegerl, an experienced diver and collector, and the team, managed to identify 88 species of live coral, 60 species of molluscs and 190 species of fish, and to explore the reef's main contours and underwater surfaces, including the area of coral sand and boulders for which the application had been made. They dropped air-force marker dye at the survey peg here and noted its spread on the currents, and they were able to take a number of photographs as well. The dives were managed from boats, and the boat-owners and crews refused to be paid for their time and work.

We certainly had our evidence that Ellison was no 'dead reef'; indeed, the divers were enthusiastic about its beauty and productivity, and they felt that with more time they would have been able to add many other species to the collection. The claim made by Mr Forbes, and by the university, was obviously more than disproved, and since Mr Forbes had made it a point in his application, his case was now undermined. The team came back to Brisbane to analyse and classify their specimens and arrange their photographs.

On 21 November, Eddie Hegerl flew up again to Innisfail, and was joined by Don McMichael. The Littoral Society's evidence stressed the importance of the algal covering of the coral sands to fish and other organisms, the vulnerability of living coral to silting, the likelihood of the fine lime silt killing Ellison Reef in earnest if blasting and dredging began. The work in the Tweed River estuary was quoted, and Eddie pointed out that the coral silt was much finer than the Tweed muds and would be carried further on tides and currents, with damage to sedentary life wherever it settled.

McMichael's evidence was mainly on the special scientific interest of Ellison; he told the court that if one unusual or rare species was found in an area, it was likely that that area might hold others as well. He contended that since one mollusc found there was known to be unique, the potentiality of Ellison Reef for further discoveries was important.

As he was a qualified marine biologist, his evidence was treated with respect by the court. He was firm that mining could not be carried out without serious damage to marine life, especially animals living in the coral sand itself; and he contended that no mining should be carried out anywhere on the Great Barrier Reef before a thorough scientific survey had determined its possible effects. Eddie, who was experienced but not qualified, had to present the Littoral Society evidence himself, as the young biologists on the dive had not been able to return to Innisfail; but in the event his lack of professional pieces of paper was made up for by the firmness and quality of his evidence and by the work done by the team on the Reef.

Dr Barnes also gave evidence. As a research consultant to the Department of Harbours and Marine and himself a member of the Great Barrier Reef Committee, as well as a medical research officer, his views too carried weight. He had been further investigating the causes of the ciguatera outbreaks and was of the opinion that mining of the Reef in general, with its exposure of bare rock surfaces, would increase the incidence of illness through eating fish that had accumulated algal poisons. He added that the Crown of Thorns outbreak,[2] which was already causing a great deal of concern, almost certainly resulted from human interference with the Reef. John, too, gave his evidence, and at the end of the case they all felt that an unbreakable case for conservation had been presented.

Perhaps as important as the evidence itself was the public interest we had managed to create. John and I had gone to the newly opened office of *The Australian* newspaper in Brisbane, and had managed to convince a young reporter, Barry Wain, that this was going to be news, and big news. We had explained our view that this was just the beginning of a major story; that the oil-drilling permits that had already been issued would cause much more controversy even than limestone dredging, and that the outcome of this case would have a great bearing on public concern for the Reef's future. Barry Wain had joined the diving team, and his stories were featured by his paper. He went to Innisfail for the case; so did reporters from other papers.

We got our publicity. The case for the Reef's future, the dangers of exploiting it, the need for research, the threats ahead, the Crown of Thorns plague − all were star news. The case closed in something of a blaze of public interest.

But there were two points which were notably against us. The Department of Harbours and Marine, which had, according to the Queensland government, responsibility for protecting the Reef, had not objected; and the Great Barrier Reef Committee had done no more than lodge a written statement with the Mining Warden, the contents of which were not made known. The Innisfail branch of the society, the Littoral Society of Queensland, and the Australian Conservation Foundation, were not even, by mining law, interested parties. We owned no property that could be damaged by mining, and we had no real right under mining law to object; public interest was no plea in law. John Büsst had financed most of the operation, with the help of a few donors and of the organisations that provided the transport and some of the equipment; he, and we, had spent much on phone calls,

[2] The Crown of Thorns starfish (*Aranthaster plated*) feeds on corals. A remarkable increase in its numbers on a few reefs was noted by scientists early in the 1960s. Dr Endean's first research into the starfish populations began in 1966; he presented a report to the state government in June 1965, with addenda in early 1969. It urged the prohibition of the taking of the triton shell, an observed predator of the starfish, and hand harvesting or other control methods. Full accounts of the Crown of Thorns controversy and government attitudes can be found in Theo Brown's book, *Crown of Thorns* (Angus & Robertson, 1972), and in *Requiem for the Reef*, by Peter James (Brisbane, 1976).

13

correspondence, air fares, and time. But we had no authority, and no responsibility; and no precedent, either. The Mining Warden had taken the opposition evidence into account, but he did not really have to do this. Everything would depend not on his recommendation, but on the attitude of the Queensland Mines Department.

Nevertheless, his recommendation was going to be crucial. We waited for it well into December. I was in the middle of the Pacific, on the beginning of a few months' travel abroad, when I had a cable from Queensland; he had recommended against mining Ellison Reef.

It was a long time – nearly six months – before I heard the final decision. The Queensland Mines Minister, Camm, had finally refused the application. The Ellison Reef case was over; a precedent had been established, not for mining the Great Barrier Reef, but for *not* mining it.

2

The Government, Science and the Reef

To understand the story of the Reef and of our various encounters with its legislative and administrative problems, we needed to look at the whole history of Commonwealth and state involvement in its administration. There was a complicated entanglement of both state and Commonwealth legislation applying to the Barrier Reef, and of the authorities delegated to administer that legislation, though no Commonwealth legislation applied to the Reef region alone.

The first instrument that affected the Reef directly, of course, was the Federal Constitution itself; and in this, offshore ownership was never made clear.

We were naturally interested in this question, since our attempt to challenge state ownership had been brushed aside in the Ellison Reef case. Over the years, John Büsst, and we in Brisbane, consulted several firms of solicitors in an attempt to find out what the legal situation in Reef waters really was.

There were various precedents in English law for the view that Queensland's boundaries, and those of the other states, ended at low-water mark. But did even the Commonwealth have jurisdiction beyond that limit? It seemed that it had simply been assumed that rights beyond low-water mark were vested either in the Common-wealth, or – if the low-water mark was not accepted – in the state, to the three-mile limit. The state had always taken the view that Queensland had rights over Barrier Reef waters, and had legislated accordingly, with no challenge from the Federal Government. But as far as our solicitors could see, there did not seem to be anything in the Statute of Westminster which would change the status as to low-water mark.

An opinion of one firm of solicitors was that there was a question whether residual rights remaining in the Crown, at the time of Federation, had in fact ever been handed over either to the Commonwealth or the states. If this was so, then legally the United Kingdom – and not Australia at all – might be the governing power entitled to grant or withhold oil-drilling and mining permits beyond low-water mark. Another opinion was that in fact the governing power might be New South Wales – whose rights, whatever they were, did not seem to have been handed over either to Queensland on separation or to the Commonwealth.

We sought out this information later in the story, when it appeared that a constitutional challenge might be the only avenue open to us. I give it here only as demonstrating the complete confusion even in legal minds at that time over the offshore ownership situation, and over the status of the actual legislation passed by the state and Commonwealth which applied to the Barrier Reef and its waters.

The Fauna Conservation Act of 1952 (Queensland legislation) established all Queensland islands as Fauna Sanctuaries. The backing for this came from an early Act passed by the Commonwealth, the Island Territories Act. Under this, offshore islands were declared part of the states, down to high-water mark. A number of islands were also declared as national parks under the Forestry Act of 1959 and subsequent Regulations, and an Amendment Act.

Queensland mining legislation allowed mineral leases to be granted on islands, except those declared as national parks. But national parks could have areas excised under an Order-in-Council. Dredging leases, under the same legislation, could be issued below high-water mark.

The State Fisheries Act 1957–62 protected living coral on the Reef. This Act was administered by the Fisheries Branch, then in the Department of Primary Industries. The fringing reefs of Green Island and Heron Island, and Wistari Reef, were also protected under this legislation, as far as their marine life was concerned, except for line-fishing of unprotected species.

In April 1971, a Forestry Act Amendment extended the jurisdiction of the Forestry Department (which then administered national parks) below the high-water mark, for the purpose of declaring marine national parks. This allowed for the protection of bottom-dwelling organisms, apart from the already-protected coral, but commercial and line-fishing were not excluded from the parks. As the Great Barrier Reef Committee pointed out in evidence to the House of Representatives Select Committee on Wildlife Conservation, in 1972, this meant that fishing in waters above declared marine national parks would still be administered under the Fisheries Act, and (said the Committee in its evidence) 'by allowing commercial fishing no protection for the total environment is available to provide the necessary baseline areas for scientific work and refuge areas'.

The Act precluded mining in marine national parks (though, again, areas could be excised), but this did not extend to petroleum drilling. The Mines Department and the Fisheries Branch had to give their views – in effect a power of veto – on the declaration of any area as a marine national park. Clearly, then, a marine national park declared under Queensland legislation would in effect have little if any protection except against the taking of bottom-dwelling organisms.

The Petroleum (Submerged Lands) Act of 1967 purported to give Queensland certain rights to grant exploration permits, production licences and pipeline licences in offshore areas. This was 'mirror legislation' with the Commonwealth's own Act, and depended on Commonwealth ratification of permits and licences. The crucial Preamble to this Act read:

Whereas in accordance with international law Australia as a coastal state has sovereign rights over the continental shelf beyond the limits of Australian territorial waters for the purpose of exploring it and exploiting its natural resources:

And whereas Australia is a party to the Convention on the Continental Shelf signed at Geneva on the twenty-ninth day of April, One thousand nine hundred and fifty-eight, in which those rights are defined:

And whereas the exploration for and the exploitation of petroleum resources of submerged lands adjacent to the Australian coast would be encouraged by the adoption of legislative measures applying uniformly to the continental shelf and to the seabed and subsoil beneath territorial waters:

And whereas the Governments of the Commonwealth and of the States have decided, in the national interest, that, *without raising questions concerning and without derogating from their respective constitutional powers, they should co-operate for the purpose of ensuring the legal effectiveness of authorities to explore for or to exploit the petroleum resources of those submerged lands:* [My italics]

And whereas the Governments of the Commonwealth and of the States have accordingly agreed to submit to their respective Parliaments legislation relating both to the continental shelf and to the seabed and sub-soil beneath territorial waters and have also agreed to co-operate in the administration of that legislation:

Be it therefore enacted ... etc.

By the decision not to 'raise questions concerning' the respective constitutional powers of the states and the Commonwealth on the continental shelf, and by later passing legislation (the Joint Offshore Agreement Act) to validate licences issued under the states' legislation in return for shared royalties, the Commonwealth had, in effect, set a trap for itself which was later to prove highly embarrassing, particularly in the case of the Great Barrier Reef. If the then Prime Minister, Harold Holt, had foreseen what problems would arise out of the legislation, it might never have been passed. It was to stand in the way, later, of proposed legislation concerning minerals exploitation offshore a problem still unsettled – and to raise other thorny questions. Its effectual recognition of the states as having offshore rights was to be a potent weapon in the fight for authority over the territorial waters of Australia, and its implications and repercussions were to be a considerable factor in the bringing down of the Prime Minister who succeeded Mr Holt, and in the problems of all future Commonwealth Governments vis-a-vis the governments of the states.

The Commonwealth's other offshore legislation included the Fisheries Act, which did not extend beyond the twelve-mile limit as far as control over foreign nationals and vessels was concerned, but controlled Australian nationals and fisheries in 'Australian-proclaimed waters' – which included waters that sometimes extended up to 200 miles offshore. The Pearl Oyster Fisheries Act, 1952, covered Australian-proclaimed waters over the whole continental shelf, and concerned pearl-shell, trochus, beche-de-mer and green snail fisheries. Australia did not take up its responsibilities for living natural resources on the continental shelf, conferred by its signing the Geneva Agreement, until it passed the Continental Shelf (Living Natural Resources) Act at the end of 1968. It then became the authority concerned with protecting 'sedentary' species on the Great Barrier Reef (and of course elsewhere.)

This is potentially a very powerful legal weapon, since the Governor-General was empowered to proclaim species as 'sedentary'. Under it, limestone mining, for instance, could easily be controlled by a willing

Commonwealth Government, since the term 'dead', when applied to a reef or to coral, is generally a misnomer. Detached underwater coral boulders and branches are quickly covered by algae and may be the homes of many small organisms of numerous species, as well as rich breeding areas for free-swimming organisms such as fish. Theoretically, then, the Commonwealth Government could under this Act protect, say, algae against removal.

But there was no overall protection or legislation whatever for the Great Barrier Reef province as such, and no official organisation had the Reef as its sole concern.

The only independent body specifically concerned with the Reef and with that alone was the Great Barrier Reef Committee. This Committee is comparatively small, with a maximum of 120 members, and is mainly made up of scientists and a few others who have worked on the Reef. Its aims are stated as being 'to promote and conduct scientific inquiry into the fauna, flora and genesis of the Great Barrier Reef; and to protect and conserve the Reef, and to determine, report on and advise of its proper utilisation'. Set up in 1922, it has been concerned ever since with the Reef and its problems.

The Committee advised the state government on many matters concerning the development of the Reef, but it did not have any official status, and the government was free to reject many of its recommendations, as it subsequently did.

The Committee's finances were never large and never reliable. It had no funds of its own beyond what it could raise for specific projects, such as the building of its research station at Heron Island. This project began in 1948, but actual building on Heron Island did not start until three years later. For this, it had a state government matching grant up to $7,000, and thirteen years later, after affiliation with the University of Queensland, a grant through the Australian Universities Commission which attracted a matching grant from the Rockefeller Foundation. The Queensland government gave it an annual grant of $2,000 to pay a maintenance officer for the station, and the operating funds for the station itself came from bench fees and grants from Australian Universities, with a Director appointed by the University of Queensland.'[3]

The Committee was far too poor to finance urgently needed research, and it depended on a very small operating fund even to run the research station it had built. Thus, its hopes were naturally centred on getting money for urgently needed equipment and maintenance for the station, and since its inception it had applied both to the state and

[3] Evidence given by the Great Barrier Reef Committee to the Select Committee on Wildlife Conservation, 1972.

Commonwealth Governments, and to overseas foundations, with very little result. It was run by honorary office-bearers, its own running costs being paid over the years since 1964 by those Australian universities which had agreed to become institutional members of the Committee. Nevertheless, the Great Barrier Reef Committee was the only authoritative spokesman on matters of Reef administration and Reef biology. Many scientists, from Australia and overseas, had used the research station at Heron Island as a base for their own research, and were members of the Committee. But since very little money was available for marine biological research, and since geological research (largely funded by oil companies and the like) was much more advanced, there were a number of geologist members who took rather different views of the Reef's potential and needs than did the biologists. And, as we were to find, even the biologists did not always agree among themselves on what the problems were and what was to be done about them.

Indeed, the Reef presents some special aspects which make it unique, both biologically and from the point of view of any possible planning and administration. To begin with, it is the largest marine coral ecosystem in the world, a 'stable mature marine community' of a very special kind. It is what is known as a 'climax ecosystem', in which hundreds of thousands of species interact in ways that are not only not yet understood, but probably beyond present scientific capacity to understand. Ecology, as against the older taxonomic and classificatory biology, is a very new appearance on the scientific scene, and there were, and are, few authoritative ecological studies even of one single subsystem such as the fish or the corals of the Reef, let alone of the Reef as a whole. The long starvation of marine biological research has meant that even these early attempts to study Reef ecology were unproven, problematic and incomplete. And biologists whose interests were not specifically in the young and struggling science of ecology tended to regard them with some suspicion, which scarcely helped us.

Another of the Reef's unique aspects is its sheer size. It runs 1,200 miles from north to south, sometimes approaching very close to the mainland, sometimes at a considerable distance from it, with outlier reefs sometimes separated by miles of deep water from other reefs and from the main outer barrier. Its islands and cays run into thousands. Many parts of the Reef are permanently submerged; others approach or reach low-tide levels. Cays and coral sand-bars are the main exposed areas on the outer Barrier, but near the shore the lovely volcanic islands themselves have fringing reefs.

To talk of the Reef, then, is to talk of many hundreds or thousands of reefs; yet it is also to speak of what is now being increasingly recognised as an ecological unity. The marine flora and fauna change in composition of species from north to south and also from east to west, and no one knows how their colonisation really takes place, or its sources, because of the complexity of the currents that carry the replenishing plankton from place to place. Biologists now often talk of the Reef as only the main system of an overall system of reefs throughout the whole Indo-Pacific region, and suspect that there may be interconnection of all these reefs through the planktonic movement across the ocean.

The Reef cannot be thought of, either, as separate from the mainland coasts, with their many fringes of great mangrove forests that form a tremendously fertile breeding-ground for many species which during part of their lives may enter the waters of the Reef proper. The interlocking and interdependent physical factors which have so long kept the Reef alive and growing, such as water temperatures, freshwater replenishment from streams and estuaries, the tidal movements which bring deep ocean water in and out of the calmer and narrower waters within the Barrier, and the winds and weather systems, are probably all indispensable to the maintenance and dynamics of its living species. These may change in unforeseeable ways when, say, a channel is blasted or a boat-harbour built. (The whole tidal-water circulation of the single fringing reef of Heron Island altered when the boat-harbour was put in, and this in turn meant that, even without the silting of the reef that followed this interference, the whole ecosystem of that fringing reef was probably also altered.)

These physical factors and their interaction are little enough known. When they are combined with the biological factors, the minuteness of scientific knowledge of what makes and keeps the Reef a living self-maintaining whole is something for awe. In its evidence to the House of Representatives Select Committee on Wildlife Conservation, the Great Barrier Reef Committee said that 'the number of species that are involved in this whole reef system are greater than any other ecosystem that we know of', and that of this uniquely large number of species, there is scarcely one whose full life-history is known.

Some think that the physical and biological dynamics of the Reef are so little understood that we may find that any interference of any kind may set off a completely unexpected reaction, whose final causes, course and total effects themselves cannot be known. Perhaps the great increase in the Crown of Thorns starfish population is an instance of this.

However, the sheer size of the Reef, and the lack of research that pointed to its possible vulnerability, meant that at the beginning of our campaign few saw the Reef as being under threat at all from developments such as mining and oil-drilling. Nor, for that matter, did many biologists.

But even the biologists of the Great Barrier Reef Committee were taken aback when they realised the extent of the permit applications and the possibilities that lay ahead for the Reef.

Although the Committee's aims included the protection and conservation of the Reef, it was not a conservationist body as such, but an organisation of scientists. Dr Endean, then Chairman of the Committee, later said that he had warned the state government of the likely opposition of conservationists to oil-drilling on the Reef; but otherwise, except in private approaches, the Committee did not openly oppose the granting of permits. For unprofessional conservation organisations like the Wildlife Preservation Society of Queensland, this lack of opposition by scientists made our position awkward. If biologists were not saying that they feared damage to the Reef from establishing an oil-industry, and did not oppose the limestone-mining application over Ellison Reef, through their own research and conservation body, it left us in an exposed position. It is worth considering why, in its early attitudes, the Committee kept silent.

For one thing, there was the Committee's strong component of geologist members who had special interests in seeing the Reef drilled – it would add to their geological knowledge, as they candidly said. For another, the ecology of the Reef was so little known that the biologists, at least those whose work had been mainly taxonomic rather than on the ecological dynamics of the Reef, would have had difficulty in supporting an argument on the basis of their own work. For a third, the controversy promised to be highly political, and scientists are generally not anxious to enter such arguments.

But there were a few marine ecologists who had worked on the Reef, or were then doing so. They included Dr Fred Grassle, now of the Woods Hole Oceanographic Institution in USA but then working in Queensland, Professor Joe Connell of California, also then working on the Reef, and Howard Choat, then a postgraduate research student working on fish biology on the Reef but now also in the United States. Their work was still in progress and its conclusions were not yet scientifically proven by publication and criticism – two factors which scientists regard as essential for establishing the accuracy of their work.

Moreover, ecological knowledge of coral reefs was scarce not only in Australia, but overseas. Even the Florida reefs had been so little studied that the aquatic editor of the journal of the Ecological Society in America, Professor Frank, in a letter to the Great Barrier Reef Committee, had to admit: 'Unfortunately we simply do not, at this stage, have major studies on the dynamic interactions between the organisms of the highly complex coral community.' He added parenthetically that 'the lack of bibliography by Australian workers clearly points to the shocking lack of appreciation by Australian granting agencies of the importance of the Great Barrier Reef as a major biological phenomenon, the prime example of a marine mature community whose study should begin to provide answers to questions of stability and diversity'.

Dr Patricia Mather, secretary of the Great Barrier Reef Committee, in quoting this letter in evidence to the Select Committee on Wildlife Conservation, added that it 'also demonstrates the lack of facilities and opportunities for scientists to work in this area'.

All these factors, then, meant that most biologists preferred, in the early stages of the Reef battle, to take the cautious stand that any operations such as drilling and mining would have to be 'controlled'.

But we did have support, and very essential and strong support, from individual biologists. Dr Grassle, Professor Connell, Dr Frank Talbot (then director of the Australian Museum), and Howard Choat all gave us information we could not have done without, expressed themselves openly and energetically on the need for protection of the Reef, and became our faithful friends. Working as we were with the Littoral Society of Queensland, of which some of them were members, we of the Wildlife Preservation Society learned much. But some of these biologists were in Australia on their own overseas-financed projects, and some returned thence during the progress of the battle or obtained overseas appointments while it was going on.

As for ourselves, 'vocal conservationists' were then a new appearance in the Australian field. We were opposing wealthy interests, entrenched government policies, and political forces that seemed immovable. 'Progress' was the watchword of the day, indeed almost a religion. If Australians had ever developed a religion of their own, it would certainly have had Mammon as its divinity. It was not until the early 1970s that the tide began to run in our favor; even now priorities have not changed.

To that extent we were out of tune with our time, a fringe phenomenon, a few societies organised on a loose voluntary basis, without finance except for what we could raise from those who had similar views, without governmental recognition, and (since 'fringe' societies often attract 'fringe' people) often accused of being composed of cranks – and cranks are not regarded tolerantly in conformist Australia. For that reason, too, no doubt many scientists did not want to be associated with us in the public mind.

Dr Bob Endean had for a time been a member of our Council,

advising us on matters like pressing for turtle protection. But when the Ellison Reef controversy began, he naturally withdrew. He was chairman of the Great Barrier Reef Committee, and its policy on Ellison Reef differed radically from ours. The fact that the Littoral Society was formed at that time, and that its young biologists cooperated with us, was a godsend.

As for the attitude of the Committee over Ellison Reef, and its view that 'controlled exploitation' of the Reef itself was not only permissible but inevitable, it seems clear that the Committee mainly wanted to get scientific evidence on the effects of limestone mining on one small separate reef such as Ellison. They certainly did not want such effects in the region of their own research station, where the work on Wistari Reef and other reefs was going on.

But, as some admitted in hindsight, they were naive. Our arguments that it was highly dangerous for the Great Barrier Reef's future to establish a precedent for limestone mining, and that Mines Departments were not to be trusted in the face of pressures to extend mining licences, did not, for some reason, weigh with them. We suspected that some kind of assurance might have been given them that the granting of a limestone mining licence over Ellison Reef would not be made a precedent. But we had tangled with the Mines Department and the state government before, and were even then working against the granting of mining licences over the Cooloola area, north of Noosa, and were a good deal more cynical in our views. We knew that to mining interests the establishment of a precedent in one area was a virtual guarantee that mining could go ahead in others. It was no part of the brief either of the mining industry or of the Mines Department to protect the living resources of the Reef.

Dr Endean, however, did draw our attention privately to the extent of the oil-drilling permits. At that time no details had been published, and very few people outside the state government and the petroleum industry had much inkling of what was really proposed.

But we did occasionally wish that the GBR Committee was more willing to accept the responsibility of its expressed aim to 'protect and conserve' the Reef. For the fact was that practically all marine biologists of standing in Australia, and with experience on the Reef itself, belonged to the Committee. This fact also rather tied the hands of the Australian Academy of Science in its entering the debate.

The Academy was interested in setting up its own committee for Reef research, and in fact made several approaches to both state and federal authorities for finance to do so. But none was given. If the

Commonwealth had set up a research committee of the Academy – which was a national body – this would in effect have over-ridden the GBR Committee's own research recommendations to the state government, and asserted Commonwealth rights over the Reef, at a time when the whole question of offshore authority was unsettled. And the Commonwealth Government was no doubt reluctant to enter the situation.

In 1961, the Committee had recommended to the Queensland government that it should set up an investigation, including marine biological research, into the Reef, in order to plan its development. The letter had apparently been pigeonholed though, as was later made clear, the state government had indeed made representations to the Federal Government for the setting up of an advisory committee – of which more later.

It was the Ellison Reef case that set off further concern on the part of the Committee. The fact that there had been no questioning of the state's right to grant such mining permits in offshore waters, and that our attempt to raise the problem of offshore ownership had been given no hearing, obviously opened up the whole question of the need for control and planning of Reef use and of authority to do this. After the Mining Warden's decision that Ellison Reef should not be mined, the Committee made a fresh approach to the state government on its earlier recommendation.

The Mines Minister was still considering whether to approve or override the Warden's recommendation. This was the Committee's opportunity to urge once again its case for the setting up of an official advisory body on the Reef. The Mines Department, apparently disconcerted by the publicity over the Ellison Reef case, was considering a survey of the whole Reef area, to provide basic information for planning and regulations for Reef exploitation. It had evidently begun to recognise the need to have scientific backing for granting – or, problematically, withholding – mining permits, in case other applications ran into the same strong opposition from conservationists. The Mines Department, in fact, did invite the Great Barrier Reef Committee to apply for funds for such a survey; and the Academy of Sciences itself made a proposal to conduct a survey.

Whatever the story behind it, the money was not granted. Instead, the state government took other measures – the commissioning of the Ladd Report, which will be mentioned later.

Of all these moves behind the scenes, of course, we knew nothing at the time. Certainly, the Great Barrier Reef Committee was disillusioned by the refusal of moneys for the scientific survey. When, with the Ladd Report in hand, the Premier announced at the end of 1968 that the state government would set up a 'Great Barrier Reef Advisory Committee' on the use of the Reef, which would advise the state – not the Commonwealth – the Great Barrier Reef Committee pointed out that the Advisory Committee could not be effective as a state body alone, without Commonwealth support. Also, it asked that any advisory committee should have 'appropriate technical experts' on the funding of research, developmental work and conservation measures, and on the granting of government subsidies to 'those bodies whose activities and involvement is demonstrated by their constitutional aims, their membership and achievements, to ensure that expert and independent advice will always be available to the Government'. Clearly, the Great Barrier Reef Committee itself was one such body.

The whole question of offshore ownership must have considerably concerned the Committee, as it was itself a state body. It was for this reason that the Committee finally decided that nothing would meet the Reef's particular case except the setting up of a statutory authority or Commission, by joint Commonwealth and state legislation. Dr Mather drew up a framework for this, and presented it later successively to the Royal Commissions on Oil-drilling in Great Barrier Reef waters, to the House of Representatives Select Committee on Wildlife Conservation, and to the Committee of Inquiry into the National Estate.

But that was to cover four years, from 1970 to 1974; and meanwhile no action was taken to control exploitation, to plan the Reef's uses and development, or to set up a body for the purpose, advisory or otherwise. It was not until early in 1975 that legislation was passed by the Whitlam government to enable the setting up of such a body; and not until mid-1976 that the first appointments to the Great Barrier Reef Marine Park Authority were announced.

In the years between, we of the conservation societies were involved in day-to-day events. And from 1968 onwards, these became more and more bewildering, demanding and exciting.

3

Move and Counter-move

In 1968 we were certain that the state government was determined to go ahead with the granting of oil-drilling permits. The Commonwealth was our only hope, and its Prime Minister was taking real interest in the questions John Büsst was raising over its responsibility and jurisdiction in Reef waters.

But we had no sooner won our case on Ellison Reef, and begun to consider the next moves, than Harold Holt drowned in the colder waters off the Victorian coast. John was both personally saddened and bitterly disappointed. But he turned to another plan – a tremendous memorial to the only Prime Minister who had taken an interest in the problems of the Reef. Not only must the Reef be declared a marine national park, but there should be a national wilderness area in each state, fronting a marine national park, dedicated to Holt's memory. All these should be under Commonwealth control. Travelling scholarships for research into marine and other ecological problems were also part of his plan.

This dream – then visionary, now perhaps to be regarded as wise and far-seeing – would need the arousing not just of Australians, but of the world, in its favour. Alison Büsst, accompanying Dame Zara Holt overseas, was charged with plans to get in touch with Peter Scott of the World Wildlife Fund, and its president the Duke of Edinburgh, and to try to convince them of the need for fund-raising for the scheme.

But the World Wildlife Fund was already concerned in many other issues, and its budget was over-committed. In any case, the Fund had itself approached Australia for help and had got none. John had to shelve his plan for the time being and concentrate on the immediate dangers ahead.

The new Offshore Mining Act and the Submerged Lands Petroleum Bill had come up for debate in the state Parliament. It was now clear that the Great Barrier Reef was in peril, and that it was quite unprotected by any legislation. With Holt gone, it was most unlikely that the Commonwealth would intervene in any way.

I, too, was now overseas. In February 1968 Arthur Fenton, our society's secretary, and John both wrote me on the problems ahead. John's letter was urgent. 'The whole of the 1,200 miles of the Queensland coastline has been quietly leased by the Queensland government for prospecting – see Planet Metals prospectus. Stir up everyone you can while you are away – the Battle for the Great Barrier Reef is now on, in no uncertain and urgent terms. Without funds, we *cannot* fight each application as it emerges in each Mining Warden's court in each country town along the coast ... We do need international support and immediate funds.'

I switched my itinerary to include Morges in Switzerland, the headquarters of the World Wildlife Fund, and I too planned to talk to Peter Scott.

I had a courteous hearing from the then head of the World Wildlife Fund, in the unpretentious house on the lake shore near Geneva. He had not then heard of the situation of the Great Barrier Reef, and he was frankly horrified. The scientific importance and the fame of the Reef were such that he could hardly believe that any government would willingly risk these for the chance of an oil or mineral strike. But though he agreed to write to the Commonwealth Government expressing this view, and suggested that the Fund might provide help in organising a research survey, there was not much that could be done from overseas. As far as jurisdiction over the Reef was concerned, this was an internal matter for Australia, since the legal question of ownership in offshore waters had not yet been raised as an international matter. Those concerned for the safety of the Reef could only watch and hope that Australia would prove a fit guardian for it. As for funding, money was not likely to be available except perhaps for mounting an international research survey.

Meanwhile, in Australia, the Great Barrier Reef Committee had itself put forward its proposals for a biological survey of the Reef to be funded by the state government. This was not welcome to the government.

Apart from the money involved, they no doubt felt that a biological survey would come up with unacceptable conclusions. Instead, they took another course. An overseas geologist, Dr Harry Ladd, was invited from the United States and commissioned to do a brief survey of the Reef and recommend on its exploitation. He had no biological qualifications, apart from being a shell-collector.

The Ellison Reef case, with its attendant publicity, had clearly disturbed the state government – but not to the extent where it was likely to give up its plans either to mine the Reef for limestone, or to drill it for oil. The Ladd survey was intended to provide justification for both.

It can be imagined that the Great Barrier Reef Committee was not pleased by the proposal. It had another reason for dissatisfaction too. Dr Endean's report on the Crown of Thorns outbreak, which the state government itself had commissioned, had been disregarded. The report had recommended immediate action by a team of divers to collect the starfish and destroy them before they spread further, and also the prohibition of collecting triton shells, which were a known predator of the starfish. The state government preferred inaction – the less expensive alternative.

But the Crown of Thorns, too, was now big news. Was it really a threat to the Reef? Reports conflicted, scientists disagreed. Finally Dr Endean demanded that his report, till then kept under wraps by the government, be publicly released. It was published by the government, without comment.

This apathy on the part of the state government towards the biological future of the Reef over which they claimed jurisdiction must have disillusioned many scientists. Yet the Great Barrier Reef committee still took no overt stand against either mining or drilling of the Reef.

The Crown of Thorns outbreak, and the question of its importance or otherwise, was now a matter for argument among scientists themselves. Until they agreed, we felt not much could be done by private conservation societies like ourselves. But it was obvious that the other threats were indeed a danger to the Reef, and for us the issue here was clear. If those scientists who pointed to human interference as the probable cause of the starfish plague were right, further interference must surely be perilous. We envisaged the Reef dotted with oil-rigs, polluted by drilling muds and wastes, intersected by pipe-lines, crowded with supply-ships, silted by mining operations. If the scientists were not willing to oppose this, then we would.

The proprietors of the Reef's tourist resorts were getting anxious and upset. All the publicity being given to the Crown of Thorns invasion might interfere with their living. Who would want to come to a reef devastated by starfish and denuded of its famous corals? Instead of pressing for something to be done on the lines of the Endean recommendations, however, they, like the state government, chose to take advantage of the arguments between scientists to play down the importance of the plague. If there was an outbreak at all, then it was a natural cyclic occurrence in the life of the Reef and the corals would soon be back to normal.

All these contrary tugs and shoves made the scene an angry one of manoeuvre and counter-manoeuvre, claim and counter-claim. The qualifications of Dr Ladd, the short time allowed for the survey, and the government's clear bias, all added to the controversy. We would need to hurry, if the drilling proposals were to be halted.

I next visited Peter Scott among his wildfowl on the banks of the Severn. He too was deeply concerned about the Reef's likely fate. But he held out little hope of our ever getting the whole Reef declared a national park. He knew more than I did of the coming battle for the resources of the ocean. In a few years, he said, the rush for control over the seabed would make the rush for African colonies in the nineteenth century look like kindergarten play. Without some international agreement on its future, Australia would stand no chance of keeping the Reef untouched. Instead, he suggested it would be best to try to get small areas declared as soon as possible. We should not ask for too much.

I was not convinced. Our belief was clear: the Reef was not just a scattered area of coral outcrops, but a biological whole to be kept undivided. What good would be a marine national park in which pollution rode through on every tide?

Back in Australia, John agreed. He was working and thinking as hard as ever. The new Prime Minister, John Gorton, must be approached as soon as possible. There were other political problems. While Gorton was known to object to overseas ownership of Australian resources, the states thought otherwise. The question of legal jurisdiction must be cleared up. But there was much unrest in Australia with the Commonwealth Government; the Vietnam War had divided people as never before. Governments could change, and an election was ahead. John decided to play new cards.

That July, Gough Whitlam, Leader of the Opposition, came to Cairns on holiday. John took off and asked for an interview. Whitlam listened with interest and sympathy to John's clear and well-briefed presentation of his case. Mr Gorton was the next target, and it happened that he too was in the north, on Dunk Island, offshore from Bingil Bay and near John's old home, Bedarra Island. John went over to Dunk Island, and set to work again.

On that lovely island, Gorton had every inducement to be sympathetic to someone with plans for protecting the Reef. It chimed well with his own views on overseas exploitation of Australian resources, too. There might also be electoral mileage to be made over the new public interest in the Reef's fate. John wrote me that he was more than pleased with the response of both political leaders.

In August I came back from Europe by sea. I had no promises in my pocket, and some dismal forecasts; but at least I knew that the World Wildlife Fund and the International Union for the Conservation of Nature were concerned and willing to help with research, if the Commonwealth Government agreed. When I landed in Perth, I saw a car in one of the streets with a striking red sticker on its window; SAVE THE BARRIER REEF, it read.

These stickers – the first bumper-stickers to appear in Australia – had been produced by the Littoral Society, working with our own society, and had gone on sale early in August. Soon thousands of cars were wearing them, and thousands of envelopes were carrying them in miniature. Also, the Littoral Society – still not as strongly against oil-drilling as we ourselves were – had organised a petition asking the state government not to permit further oil and gas drilling on the Reef unless adequate, detailed plans had been made to deal with any oil spillages rapidly and in a manner which would not harm aquatic life. This had gained 13,000 signatures.

I went to Bingil Bay on my way back to Brisbane via Darwin, and spent a few days with the Büssts. John had a great deal of briefing for me. The Ladd Report was not yet out, but already it was being given bad publicity from the Reef's sympathisers. And John had found a new ally in the recently appointed Professor of Zoology at the University College in Townsville. Professor Burdon-Jones had a vision of a great Institute of Marine Science, to be based at Townsville. Such an Institute would be able to use the Reef, that treasure-house for scientists, as a research area. It could lead the world in investigations of marine biology, and its work could be immensely important in the coming exploitation of the oceans for food and resources. For this, the Reef would need to be protected from damaging exploitation, and here the Professor's interests and John's own were at one. Both of them had approached the

Prime Minister with this new proposal, emphasising the urgency of the need for research, and how little had yet been done.

It was a major proposal, and it would be expensive; but North Queensland would benefit, as well as the world. The Prime Minister was obviously interested, and John was most enthusiastic. The Reef as a scientific laboratory for the world! It was far better than the Reef as an oil reservoir and source of limestone.

The controversy over the Reef had set off a number of new branches of the Wildlife Preservation Society, too; we had branches at Townsville now, and at Ingham, as well as Innisfail. They were working eagerly, on the new wave of interest in conservation, and the fate of the Reef was their chief theme.

The state government was now playing a defensive line. Care for the Reef's future, they now said, was one of their chief pre-occupations. Nothing would be done that damaged it.

But the *Torrey Canyon* disaster off England's south coast, earlier in the year, had set off a clamour of publicity on the dangers of oil spills in the ocean. This worked against the oil companies' assurances of safety for the Reef, not only in oil-drilling, but in transportation of oil. The Reef's inner passage was just as endangered by the great tankers feeling their way through Torres Strait and down the coast as was the south of England. Could we demand that they take a different route?

John had been talking to experts about this; but clearly it would be very difficult. A wreck outside the Barrier would be just as damaging as one in the inner passage, and the outer route was dangerous, badly charted and without navigational lights. The alternative route by Western Australia would be extremely expensive; to advocate it, at this stage, might involve international problems. It would certainly involve fighting the Commonwealth Government, as well as the state, and at that stage this was something we did not want to do, for obvious reasons. But sitting on the Büssts' veranda, overlooking that aquamarine stretch of sea, we feared those tankers.

Still, if we were able to arouse enough concern about the Reef on the oil-drilling question – which was at least possible – we might be able to change government thinking and public opinion, and that might lead on to controls on tankers as well.

At the end of August, John wrote to the Prime Minister with full details of the prospecting leases already given out by the Queensland government – which at that stage were not public.

As from September 1967 some 80,920 square miles of the Great

Barrier Reef has been leased by the Queensland Government for oil and mineral exploitation:
(a) to Pacific-American Oil and Shell Development (Aust. Pty. Ltd.) – P. 70, 3,323 square miles.
(b) Australian Oil and Gas Corporation – P. 90, 57,000 square miles.
(c) Ampol Exploration (Qld.) Pty. Ltd. – P. 97, 8,007 square miles.
(d) Tenneco Aust. and Signal (Aust.) Petroleum Co. – P. 111, 5,340 square miles.
(e) Corbett Reef Ltd. – P. 127, 7,250 square miles.
As you know, on the discovery of oil or minerals, licences to proceed with actual mining follow automatically ...

I suggest: A complete moratorium on all mining activities for at least ten years until such time as our own, and international, scientists have had the opportunity to determine what is possible and what is definitely not possible;

The Commonwealth take full control of the area *now*,

Full financial support for the proposed research centre in tropical marine science at Townsville.

He wrote to Whitlam notifying him of these suggestions. Whitlam had already set to work in the House, where he and Dr Patterson had been asking relevant questions and emphasising the importance of the Great Barrier Reef to Australia's and the world's future.

For the first time, the Great Barrier Reef was occupying a large place in Canberra's thinking. The state government began to feel beleaguered. It happened that the responsible Mines Minister, Camm, represented the Whitsunday electorate. He went up to soothe the anxieties of his electors and of the Reef resort proprietors, and travelled through the electorate and to the islands offshore.

He said he envisaged 'four or five' marine national parks along the Great Barrier Reef and that 'an expert committee' was working on recommendations; he said that island resort proprietors 'kept a watchful eye' on areas of their islands that were already national reserves. When the marine national parks were declared, taking of fish and shells and marine specimens would be prohibited, and for that reason the whole Reef could not be declared a national park; but he felt sure the park areas would give 'full protection' to the Reef and its tourist attractions.

John wrote to *The Australian* newspaper – one which was finding its conservation angles a useful selling-point, and with which we already had a good deal of contact. He gave the facts about the mineral and oil exploration permits. He outlined the conditions under which Dr Ladd was working on the Reef report.

The Queensland Government recently imported Dr H. S. Ladd, an American geologist whose specialty is shell classification, allowed him one month only to survey the whole 1,200 miles of the Reef – a ridiculously impossible feat – and has stated that it will be largely guided by his advice on its scheme to mine millions of tons of so-called 'dead coral rubble' for the cement industry. This will undoubtedly cause a storm of international scientific protest. I propose to see that it does. There is no such thing as 'dead reef'. The so-called 'dead reef' provides the vital feeding and breeding grounds for the multiple organisms of the Reef, and in fact constitutes by far the major portion of the whole area of the Reef … The so-called 'dead reef' is in fact the very basis and living heart of the Reef.

Siltation from large scale mining for minerals, or by both pollution and siltation from oil drilling (the main characteristic of the huge oil rigs is mud) would mean large scale disaster to the Reef. Mining for limestone, for example, would not be economically feasible without large scale blasting, and this, both by pollution and by the destruction of vital and delicate food chains, could extend [silting] far beyond the original area of mining operations. Already, in Queensland alone, there are enough known deposits of lime (Bureau of Mineral Resources, 'Limestone', No. 26) to supply industry for the next 200 years, without any further exploration … It is therefore criminal folly to destroy the Reef by using it as a source for limestone or for any other minerals.

Mining is a wasting asset. Oil wells or mineral fields are inevitably limited in extent, and once exhausted will leave us with a marine Simpson's Desert where once there existed the fantastic splendour of the Great Barrier Reef. Aesthetically and scientifically, it is equivalent to bulldozing the Taj Mahal or the Pyramids for road material. 'The Reef is the largest structure on earth ever created by any organism, including man.' (Prof. L. C. Birch) The myriad separate reefs must be regarded as one huge living structure, closely interconnected to an extent which is not as yet fully scientifically understood … There is as yet no known method of controlling pollution on the Reef. The words 'controlled exploitation' are an idle and meaningless political phrase …

Once a mining licence is granted, it is a valid legal contract and cannot be rescinded. In the whole of the Oil and Petroleum (Submerged Lands) Act 1967, there is no penalty for infringement of licence greater than one thousand dollars. Even if penalties of one billion dollars were imposed, it would not enable us to replant corals …

He went on to outline the proposals he had already made to the Prime Minister. He suggested that if the Commonwealth laid immediate claim to ownership of the Great Barrier Reef, it could argue about this afterwards in international law. 'No nation at this juncture of history is liable to declare war on us if we claim the Reef. We have the full sanction of international law by virtue of our historical and continued usage of the Reef area.'

It was just the importance of the Great Barrier Reef for ecological studies which was bringing the young biologists here from overseas. Like the northern rainforests, it represented to them one of the greatest marine ecosystems of the world, where the interactions of living things could be studied in their highest complexity. Len Webb, whose rainforest ecology studies were of the same nature, understood their views as few others could.

The Brisbane headquarters of the society kept working for publicity; I wrote letters to the press and held interviews. With the Ladd report due soon, we kept up the pressure. When it appeared in early September, it confirmed our impression that as a geologist, Ladd was more interested in encouraging mining and drilling than he was in Reef protection. Since his visit was sponsored by the state government Department of Mines, which by now had very strong views on conservationists, it was not surprising that he departed from his scientific brief to attack those who had been pressing for Reef protection.

Our own views found support in the opinions of the young marine biologists. Rather than examine the Ladd Report myself, then, I reproduce the text of a letter to the press by one of them, Dr Fred Grasile.

A report by Dr H. S. Ladd has just been made available advising the Queensland Government on its policy for conservation and use of the Great Barrier Reef. There is no one more competent to advise on the mineral potential of the reef and the possibility of mineral exploitation. Dr Ladd's report might well serve to indicate what might be used but is of little help in indicating the unique features of the reef which should be conserved.

There are few places in the world so unique that they should be maintained for future generations. Some such places are the Galapagos Islands, Aldabra Island, Lake Tanganyika, Lake Baikal and The Great Barrier Reef. These are unique environments with organisms found nowhere else in the world. One of the reasons these places are biological wonders of the world is that the environment in these places has remained relatively stable over a long period of time. Most of the organisms are ill-adapted to sudden changes. Human disturbance which would have little effect in less stable areas can have catastrophic consequences in these environments.

A great deal has been made of the conflict between the 'extreme conservationists' on the one hand and the exploiters on the other. I think this conflict is a myth ... Labelling organisations other than Government Departments and the Great Barrier Reef Committee, such as the Australian Academy of Science, the Queensland Littoral Society, the University of Queensland and the Wildlife Preservation Society of Queensland as less 'broadminded' or, by implication, 'only interested in conservation', 'extreme', or 'well-intentioned', adds nothing of scientific interest to the report. Since Dr Ladd's survey was sponsored by the Queensland Department of Mines, such statements jeopardize his position as a disinterested observer ... Dr Ladd mentions that at the other extreme from the conservationists are 'individuals and some companies that look upon the reef as an untapped resource so vast that it would not be seriously damaged by mining or other activities, particularly if operations can be confined to a few accessible marginal areas where living coral is sparse or absent'. On the question of limestone mining, this view appears to differ little from that of Dr Ladd.

Dr Ladd states that 'rich coral growths that give the reefs their name are extremely limited, averaging less than one-fourth of the surface and near-surface reefs'. He also states that 'as a source of agricultural lime and limestone for cement manufacture, parts of the reef that do not now support living coral can be developed, if sites are carefully selected and operations rigidly controlled'. If these statements are interpreted uncritically, they suggest that three-quarters of the reef could be mined.

When Dr Ladd specifies how mining sites are to be chosen his only criteria are the purity and uniformity of the calcium carbonate and whether local currents will carry silt to surrounding areas resulting in coral death. Elsewhere, the report states: 'Any deposit that appears dead may be part of a reef complex that, on the whole, is very much alive. Mining of such a reef would be undesirable.' In view of this, are not Dr Ladd's criteria for suitability of mining sites irrelevant? My own view is that any area on the Great Barrier Reef, whether or not it consists of much living coral, is part of a reef complex that is very much alive! Dr Ladd might even agree with this as he says: 'As far as coral growth is concerned, the barren areas can be called dead but this does not mean that they do not support many other reef organisms or that such "dead" areas do not play important parts in reef ecology.' Why then doesn't he mention other reef organisms or reef ecology when he presents criteria for selection of areas to be mined?

My attitude on coral reefs is different from that of Dr Ladd and I should explain why. I am an ecologist doing postdoctoral research on reef communities at Heron Island. As an ecologist I am just as interested in the other organisms of the reef as I am in the living coral. During the past year I have found that dead pieces of coral contain the greatest diversity of species yet reported in any marine environment. I do not want to see these animals destroyed needlessly nor do I want the fish which depend on the animals and plants of the dead coral to be destroyed.

It is worth examining the biological evidence in Dr Ladd's report. He found living coral flourishing on the inner margin of the reef flat of the Swain Reefs, the lee areas of Heron and Wistari Reefs, and the inner edge of the Outer Barrier. Of the reefs inside the Outer Barrier he comments: 'I landed on a number of the inner reefs and with the exception of the reefs near Green Island found them all alive and healthy.' From the itinerary in the report these appear to be the only parts of the reef that Dr Ladd examined closely with the exception of some fringing reefs near the mainland and in Torres Strait. The only reefs that he found not supporting living coral are in the vicinity of Green Island. Does this mean that Dr Ladd thinks that the reef around Green Island might be mined? Would the thousands of tourists who have visited Green Island this year agree that the reef there is so dead that it is not worth preserving?

I heartily support Dr Ladd's proposal for an aerial survey of the reef but I do not think it will answer the crucial questions: (1) what kinds of exploitation result in permanent damage to the reef and (2) to what extent are different parts of the reef interrelated? Dr Ladd gives his personal view that 'it would be difficult, if not impossible, to replant a rich and varied patch' of coral. This is probably true and, in addition, it is unlikely that any of the threatened plants and animals of the reef could recover from widespread destruction during our lifetime. We do not know how irreversible changes in the reef might be. An answer to the second question would be even more difficult. In addition to the maps Dr Ladd recommends, we need detailed knowledge of the hydrography, currents, plankton, life histories of animals and, not least, which animals are there ... The answers to even the most simple questions concerning the effects of exploitation present unusually difficult biological problems ...

Before adequate answers are available about the effects of reef exploitation, research must be done into the basic functioning of coral reefs. The Great Barrier Reef is not just another coral reef – it is *the* reef – the biggest, the most varied, the most complex system of reefs in the world. Any research on coral reefs must finally be referred to the Barrier Reef since all the features which make coral reefs unique environments are best developed there. Australian scientists and scientists from all over the world would come to do this research if the government would provide even modest support specifically for coral reef research.'

Dr Grassle was not alone. Professor Burdon-Jones argued on the ABC that the state of scientific research on the Reef demanded that there should be no mining at all without major scientific research. He pointed out the need for the Townsville marine research institute and for involvement of the Australian Academy of Sciences. (This may not have pleased the Great Barrier Reef Committee.)

The Committee itself did not want to say or do anything that implied lack of faith in the state government's intentions on the Reef. But they were in something of a dilemma.

They proceeded cautiously. Dr Endean gave a press interview in which he said that to 'leave the Reef alone' would be impractical and unprogressive. 'The state government,' he added surprisingly, 'has not put a foot wrong yet.'

Our own reaction to the Ladd Report, which we knew would of course be passed on to the Commonwealth Government, was less tolerant. Soon afterwards, John Büsst came to Brisbane for a meeting, to discuss further tactics. There was a rising tide of public interest in favour of Reef protection which had to be kept going. John's public raising of the question of who really had authority to grant mining licences, and who really had authority over the Reef, was an obvious line to be pursued. It would not be welcome to the coalition parties, but we were now sure that the Prime Minister was sympathetic. The Ladd Report might not influence him.

At the University of Queensland, John interviewed scientist members of the Great Bather Reef Committee. Still understandably sore over our successful defence of Ellison Reef, and our disproving of their claim that Ellison was 'dead', they were unreceptive. They still took the view that the Reef could, and must, be exploited – provided they were given the money to do the research they wanted beforehand. Our uncompromising stand with the younger ecologists nettled them. They had chosen to accept the state government's continual assurances that they meant the Reef no harm, and to interpret this as meaning that oil-drilling and mining would not, after all, be allowed – or not on such a scale as to cause what they might regard as a serious threat to the Reef. Accordingly, our approach was not popular.

We shifted ground. The Australian Conservation Foundation's annual general meeting was to be held in Canberra in the following month (October). Their July newsletter had set out the stand they were taking:

> Recently, several independent events have shaken the complacency of Australians, who assumed that the Great Bather Reef was secure for all time against natural or man-made destruction. The first was the outbreak of a plague of the Crown of Thorns seastar, *Acanthaster planci*, which has now been found in varying concentrations on reefs to the southern Bunker group ... Research by Queensland scientists has not so far revealed either the cause of the outbreak or any means to control it.
>
> The second event was the granting of Authorities to Prospect for offshore petroleum deposits to a number of companies over much of the Barrier Reef. Already two wells have been drilled ... south-west of the Swain Reefs in the Capricorn Channel, and undoubtedly more will follow ... The third event was an application to mine for coral detritus on Ellison Reef ... The fourth event was the intrusion of foreign fishing vessels into Barrier Reef waters, and the destruction for food of large numbers of Giant Clams by their crews. This led to consideration of the rights of Australians to the fishery and other resources of the Great Barrier Reef, and the alarming discovery that Australia's title to the resources of the Reef waters and the reefs themselves was not clear ... The latter involves complex questions of international law, and poses a threat to the conservation of the Great Barrier Reef by making the enforcement of fisheries protection laws difficult.
>
> The Australian Conservation Foundation has recommended that a proper survey of all aspects of the Great Barrier Reef should be

made before any further action to permit exploitation of any kind is taken.

Two influences were clear in this article: that of scientists such as Dr McMichael, the Director, and that of the legal approach (the then President of the Foundation was Sir Garfield Barwick). As we were to find, Sir Garfield had already adjudicated in a case which raised the question of offshore ownership. We were glad of the article, and felt the Foundation's attitude should be taken further. We wanted a clear-cut policy which included not only the need for research and for legal definition, but actual protection of the Reef.

John Büsst was a member of the ACF, but not a Council member as Len Webb and I were. He evolved a form of motion to be put at the annual general meeting. It read:

> That the Commonwealth Government be asked to lay immediate claim to Australia's legal ownership of the whole Great Barrier Reef area, by virtue, in international law, of our continued and historical usage of the area, since the arrival of Captain Cook.
>
> That an immediate moratorium, against any mining whatsoever, whether for minerals or for oil, be declared over the whole Reef area, until such time as a full scientific investigation shall determine what is permissible and what is definitely not permissible by way of human interference with the natural processes of the Reef.
>
> That, as the preservation of the Great Barrier Reef is not only a matter of Australia-wide concern, but also of deep international scientific interest, the Commonwealth Government be requested to set up a Commission, under the auspices of the Australian Academy of Science, to establish both the trusteeship and the full scientific investigation of the Reef, such Commission to have full powers to obtain the advice, assistance and membership of all international scientific organisations as required.

This was the first attempt to work out how the Reef should in fact be administered, and to by-pass the accepted role of the state government. The fact that the Foundation's president was a lawyer, and that the Foundation was itself supported financially by the Commonwealth Government (being the only conservation body given any financial support at all) seemed to make the ACF the logical platform for advocating Commonwealth control of the Reef.

At that time the ACF had a Queensland arm (called the North-east Regional Committee). It had organised the symposium on Caring for

Queensland, at which Don McMichael had spoken the previous year on the need for marine national parks, and it was naturally interested in the whole question of the Great Barrier Reef. But the Foundation was fearful of allowing its regional groups any autonomy; its anxiety to be seen as a scientifically conservative body made it highly wary of the kind of campaign into which we had entered; and the members of the regional committee largely supported its views. In effect, the campaign for protection of the Barrier Reef was being run by non-scientists like John and myself, and by voluntary amateur organisations. It was a battle for publicity on the one hand, and on the other, a lobbying campaign directed at politicians. The ACF relied heavily on the low-key approach, and on the support of scientists who might themselves take alarm at such a campaign.

We sensed, then, that an attempt to harden ACF policy might not be popular. But clearly it was urgent. We planned a tactical approach.

Len and I were likely to be the only Queensland members of the Foundation's Council who would support the motion. Things were moving far and fast, and the ACF might fall into the mire of controversy (as others saw it) if the resolution were passed. The fact that even scientists were divided in their views on the Reef's problems, and on who should conduct the scientific surveys that all agreed were needed, made the Foundation's situation delicate. But we had a good hand to play. No biologist could have contended that the Ladd Report represented a 'thorough scientific survey', such as the Foundation itself had already recommended before exploitation might be permitted. It seemed likely that, unless some action was taken at Commonwealth Government level, the Queensland government would simply go ahead both with limestone mining and oil-drilling permits. it was clear that no low-key approaches or dignified protests would make any difference to this; and it was also clear that until the question of offshore ownership was settled, the Commonwealth could take no action on the Reef at all.

John was to speak to the motion, I to second it. But John had a problem; his voice had almost disappeared. The tension of the past year or two had increased his smoking; now he had throat trouble. He was to go on to Melbourne afterwards for medical examination. But when the time came he spoke strongly, and as persuasively as ever. Our weakest cards in pressing for the conservation of the whole of the Reef lay in the ecologists' arguments; however convincing they were to us, more traditional scientists distrusted them. But a number of members of the Academy of Science were present, and it may be that the attitude of the Great Barrier Reef Committee and the Queensland government to the Academy's earlier proposal for a scientific survey had weight with them. We emphasised, too, that a moratorium was absolutely necessary, if the Reef's potential importance as a scientific laboratory for research work was to be preserved.

Sir Garfield, in the chair, made the point which was later to expand into a thundercloud. It would be dangerous to 'intrude into the delicate political and legal relationships between the Commonwealth and the states as to administrative control', he said. The motion as finally amended read:

> That the unique national and scientific significance of the Great Barrier Reef requires that it, and the area it occupies, should be wholly under Australian control, and that the Commonwealth Government should lay claim internationally to all those parts of the Reef and the area it occupies which may not now be internationally recognised as under Australian control.
>
> That it would be most desirable that the whole of the Reef and the area it occupies should be under one control.
>
> That there should be a thorough scientific investigation at the highest level of the Reef and its area, to determine what if any human interference in the course of industry or commerce could be allowed without harm to any part of the Reef or the area it occupies, and that pending the results of such an investigation, the Commonwealth and state governments should ensure that no industrial or commercial activity, other than fishing and tourism, takes place on the Reef or the area it occupies.

There was some obvious dissent, but the motion was carried overwhelmingly. We were delighted and relieved. The only national and government-supported conservation body now had a policy which recognised the importance of the Reef as a whole and its protection as such and not in separate marine parks; it saw Commonwealth involvement in that control as vital; the need for recognition of this under international law was stated; and the major scientific investigation was seen as having to take place before any major exploitation of the Reef could be allowed.

John went on to Melbourne for his medical examination. The examination found a pre-cancerous condition in the larynx; he was allowed to go back to Bingil Bay for two months, but neither smoking nor talking was to be allowed. For John, at such a crucial time, the silence would be grievous.

4

Tragedy Abroad

John had been not only the first fighter in the Barrier Reef field, but our indispensable spokesman, contact with politicians, diplomat and tactician, as well as friend. We were all deeply worried over his illness. Moreover, it was not only the Barrier Reef that we were working on. We had for several years been engaged in the fight for a coastal national park at Cooloola, and for a change in the Mines Act to allow for the consideration of public interest. There was much work to be done.

In November came the news that a joint Japanese-Australian survey of the Great Barrier Reef was to be done. A small Japanese submarine, the *Yomiuri*, owned by a Japanese newspaper magnate, would carry a number of scientists through Reef waters. It was supposed to be a biological expedition, but when we heard that the survey was being partly financed by the Commonwealth Government's Bureau of Mineral Resources, we began to wonder what new factors might be taking a hand at Commonwealth level. We discovered that the magnate in question also had considerable oil interests.

On 21 November, 1968, the Commonwealth Government passed the Continental Shelf (Living Natural Resources) Act. It went some of the way towards claiming the continental shelf, and therefore the Great Barrier Reef, as an Australian domain. The Commonwealth now had responsibility for sedentary living species on the Reef, not only within the twelve-mile limit but up to the boundaries of the Reef itself.

The clam-fishing boats from Formosa and elsewhere, which had been taking many clams from the Reef, were now operating illegally. So far so good. But when we studied the Act we were dismayed to find that though living coral and other organisms were now under Commonwealth control, there was no protection for the so-called dead coral and coral sands – which were also the limestone that we knew the Bureau might well be interested in.

In his speech on the Bill, the Minister, Mr Anthony, said that the Commonwealth was 'considering' preparing legislation to deal with non-living resources apart from petroleum (which was of course covered by the joint offshore agreement with the states). Apart from silica sands, only limestone was likely to come under this classification on the Reef itself. As for the moratorium on mining that we had hoped for, there was no word on this.

John, fretting in enforced silence at Bingil Bay, worried about the submarine survey and about the new Act and what it did not do. He wrote letters to the Prime Minister, to Gough Whitlam, and to others for information, but got little.

We scanned newspapers for possible clues, and found alarming ones. The big new ironworks complex at Kwinana in Western Australia was being opened. It suffered from a big problem in materials-handling. Iron ore for the blast furnace had to come from over 300 miles away; it had to be purified, and one of the materials needed for this process was limestone, which was being brought not only from South Australia, but from Japan. And Japan was the country which would take the main export component of Kwinana's production.

We could only hope that the connection we made was wrong, and that the financing of the Japanese submarine survey by the Bureau of Mineral Resources was just a coincidence.

I wrote to tell John this news, and added that it looked as though the moratorium on mining would be much more difficult than we had hoped. The States-Commonwealth conflict was just as delicate and problem-loaded as Sir Garfield had forecast.

The submarine expedition was also charged with examining the reefs for signs of the spread of the Crown of Thorns. Dr Endean was aboard.

Though we were taking the Crown of Thorns question very seriously, scientific disputes about it were still raging, and until there was some definite agreement, we preferred to keep out of that field. What alarmed us, however, was that if 'dead reefs' were being reported in the wake of the starfish, this might be used as an argument for allowing limestone mining in such areas. And meanwhile, the Commonwealth's proposed legislation on mineral rights over the continental shelf appeared to be striking some snags. No more was heard of it.

The then new director of the Australian Conservation Foundation, Dick Piesse, arrived in Brisbane early in December to talk over conservation problems, including that of the Reef. He told us that the Commonwealth's proposed legislation on minerals was being held up, as we had suspected, by the Commonwealth-States impasse, though the idea of a moratorium was still being pursued. The proposal for a scientific survey with Academy of Science participation was held up too, and could not be financed until the legislation had been passed and the Commonwealth's claim to the control of the continental shelf was complete. There was no sign of a High Court case to decide offshore ownership.

The Australian Conservation Foundation had decided to hold a symposium on the Reef early in 1969, and we discussed what this might cover.

Marine ecologists with experience in coral-reef areas which had pollution problems were not available in Australia. If possible, we wanted one brought to Australia as one of the speakers. We were not only concerned with possible pollution from mining and oil-drilling, but with the effects of pollution from coastal rivers. We knew that Queensland's rivers were already carrying heavy pollution loads. This had already been pointed to as one possible cause of the increase in Crown of Thorns starfish, for starfish were thought to be very resistant to pollution, while their predators might not be so.

The Great Barrier Reef Committee held a full meeting of its members at this time, and some very good news came from this. The Committee had now reversed its views. It agreed that no mining or drilling should take place before the full-scale scientific survey had been completed. This agreed with the resolution passed in October by the Australian Conservation Foundation, and meant that all scientific bodies in Australia with an interest in the Reef were now at one on the need both for a moratorium and a scientific survey.

Our expedition to Canberra in October had been fully justified, and John was delighted with the news.

Now began the first of a series of accidents involving oil which were to have a lot of influence on public thinking. There had been so much publicity over the *Torrey Canyon* wreck off the Cornish coast, and its results, that the dangers to the coast and seas from oil spills and blowouts were in the limelight. Now there was a big gas blowout from one of the Esso–BHP oil wells in Bass Strait – the second such incident since 1967. What effect was there on marine organisms?

We tried to find out through the Victorian Fisheries division, but information was hard to come by, even for the 1967 accident. There had been reports then that the coast of Victoria was littered with dead fish and that fish being taken by professional fishermen offshore were blind.

Now John had an answer to one of his numerous letters on the question of limestone mining and the actual value of 'dead coral' in reef ecosystems. He had written to the Institute of Marine Sciences at the University of Miami asking for an opinion. Professor Lowell Thomas had now replied.

It is a curious fact that the general public and some uninformed scientists consider that a reef without living coral is a 'dead reef' … The folly of such a consideration may be seen when one realises that almost the entire surface of a 'dead reef' is in fact living. The term 'dead reef' is a misnomer. It refers to portions of a reef where there is little or no living coral. Such areas are covered by coral rubble, boulders, and shingles of various shapes and sizes. Most of this rubble is coated by living coralline algae. In addition to the coralline algae, filamentous algae is usually very common in these 'dead reef' areas.

To survey the worms, shrimps, small crabs, fish, anemones, snails, clams, brittle stars, sea–urchins and other organisms that live in such 'dead reef' areas would in itself be a formidable task. One usually finds many of the same organisms living in these areas that live in the 'live reef' areas. There is a basic difference between the two habitats, however, in that many of the larger fish which use the 'live reef' as a hiding place during the day may travel long distances at night to feed on 'dead reefs' and grass beds … Biologists are beginning to understand that the 'live coral reef is far from being a self-contained ecosystem. Such a coral reef' is only part of an ecological continuum which includes adjacent areas that may stretch for many miles … Thus the food cycle of the reef involves areas many miles distant, and water that may have travelled for tens or even hundreds of miles. The maintenance of a 'live reef' fauna depends upon proximity of grass beds and 'dead reef areas'.

This not only confirmed what the marine ecologists had told us, but went further. Clearly the whole Reef province had to be viewed as an ecological totality, not as a series of areas some of which might be exploited without affecting the rest. Publishing the letter in our news-sheet, we pointed out: 'This confirms even more strongly our view that mining of any kind on the Reef may well finally result in uncontrollable and widespread damage to areas far beyond those in which the actual mining takes place. The Queensland government has now called tenders for oil-drilling offshore.'

It had; and the press reports indicated how much competition there was for them. Forty groups of oil companies were moving for drilling rights, in the Gulf of Papua and off the Queensland coast. Many of them were backed by some of the biggest international oil search companies. The northern area between Papua and Queensland was of about 1,800 square miles; the next to the south, about 1,150 square miles, and the lease from near Maryborough to the Queensland border was about 5,900 miles. All the blocks were either wholly, or in part, within a 200-metre depth limit; and most of them stretched well beyond the three-mile territorial limit.

Among the lease areas was one in Princess Charlotte Bay, on Cape York, and the applicant who succeeded in gaining this was Exoil-Transoil, the company in which the Premier, Mr Bjelke-Petersen, held so strong an interest.

There was immediate question over the granting of these huge areas. The state Mining Engineer replied that there would be 'adequate safeguards' to protect the Great Barrier Reef from effects of drilling. He did not explain what these safeguards were, but said the regulations[4] were being drawn up by the Mines Department. Queensland would take sixty per cent of the royalties, the Commonwealth the remaining forty per cent. These royalties would be based on ten per cent of the well head value.

With the Esso-BHP blowout in mind, the public began to ask questions. The Premier of Queensland insisted that the 'most stringent safeguards' would protect the Great Barrier Reef, but again he did not specify what these were.

Now we were collecting documents on effects of oil pollution on marine animals. Since the Commonwealth was now controlling 'the living resources' of the Reef area and was responsible for protecting them, we had to get as much information as we could, but scientific articles with a real basis of research were not easy to come by.

We found the journal of the American Society of Mammalogists,

---

[4] It emerged in evidence to the Royal Commission later that these regulations were still in draft form only when drilling was due to begin, and were not in force; thus the only regulations that would have applied to the Ampol-Japex drilling, at least, were those applicable to drilling on land. The draft had other highly unsatisfactory aspects.

published in November 1968. This gave us precious information on the results of oil spills in Cook Inlet, Alaska. We used this as a basis for a letter to the press.

The Queensland Government has recently called tenders for extensive oil search drilling rights on parts of the Gulf of Papua and the Queensland coast.

The State mining engineer, Mr I. W. Morley, has stated that there will be 'adequate safeguards to protect the Great Barrier Reef from the effects of drilling'.

Can he, however, give any assurance either that oil spills from such off-shore drilling will not take place, or that, if and when they do, such spills will not be allowed to harm the reef?

The American Society of Mammalogists is extremely concerned over the results of oil spills from drilling in Cook Inlet, Alaska. Since the middle of 1966, oil-spill 'incidents' from drilling rigs there have amounted to no less than 75.

Large-scale mortality of marine life has resulted. It is stated (*Journal of Mammalogy*, November 1968) that one spill alone killed 1,800 to 2,000 ducks. Bird corpses are comparatively easily counted; mortality of marine life and plants is not mentioned.

The results of such 'incidents' on marine life in shallow tidal areas (which are highly important in fish-breeding) would certainly be even more disastrous.

There can be no assurance that oil pollution by drilling rigs can ever be wholly prevented, however carefully the operations are carried out; and there is at present no known method of removing petroleum products quickly and efficiently from marine spills. Detergents are even more harmful than the oil itself.

There is one method, however, which should be explored: that of insisting on slant-drilling from inland wherever this is at all practicable.

The Society of Mammalogists has requested in the case of future Alaskan drillings that this method be encouraged, and that drilling from oil rigs erected over the water should be completely prohibited until methods of preventing oil spills are developed and until petroleum can be 'quickly and efficiently removed' before damaging animal and plant life.

The reef is not just a 'scenic area' for tourists, however important this aspect may be. It is one of the greatest, perhaps *the* greatest, biologically productive area in the world today.

As a tremendous food-producing and scientific asset, and one highly vulnerable to human interference, we should be ensuring its safety by every possible means.

We have recently been warned that Australia is polluting its coastal and inland waters at a rate dangerous for our own future.

We appeal to all interested Australians and to all Government members, both State and Federal, to be alerted to these dangers and to be ready to take immediate action before the reef is jeopardised beyond repair.

We were managing to keep up the publicity and therefore the public interest, but there were still people who were inclined to believe the state government's assurances that no harm would come from drilling the Reef, and that a blowout was highly unlikely. An unfortunate statement by a Queensland member of the ACF, in a press interview, gave the public the impression that the Foundation would not have any objection to drilling, provided a scientific survey had been made. This was specially ill-timed, for at a press conference in Sydney during December the Prime Minister had for the first time publicly indicated that he himself would like to stop drilling and mining on the Reef. We flooded ACF headquarters with protests. Would they publicly contradict the impression the report had given?

But once again the ACF dragged its feet. No public denial was made. The ACF was preparing the ground for the May symposium on the Reef; and, though we did not know this at the time, they were arranging for representative speakers from the oil and mining interests, in line with their policy of 'keeping the balance'. No doubt a too-strong statement of what, after all, was their declared policy might have interfered with the even-handed image the ACF was trying to present.

In fact, they were concentrating their hopes for the symposium not on the biological arguments against mining and drilling, but on the legal aspect. Who actually had authority over offshore areas in the matter of the Reef's possible oil and mineral resources? The Commonwealth's legislation depended on the answer. Since, over the years since Federation, it had left practically all decisions on the use of the Reef to the Queensland government, the answer would be crucial.

Dr Ratcliffe, replying to my protest over the Press interview, explained this, and enclosed a copy of the draft resolution which he hoped would be passed at the symposium. It leant heavily on the legal aspect of ownership, but it took little account of what seemed to us to be the burning question that needed publicity – that of the damage to the Reef's living resources that might come from mining and drilling, and of the need for a moratorium on both. We criticised the resolution and suggested alterations and additions, but in the event it was to be presented to the symposium just as it stood in the draft.

John was due in Melbourne for his next medical examination in January. Unable to do much talking, he had been very busy with his pen, working both on politicians and on local party branches. The announcement of the extent of the tenders for oil-drilling rights had caused a great deal of public disquiet, and politicians were beginning to worry over this. He used Professor Thomas's letter to good effect.

In Brisbane, the strain was beginning to tell. We had all been working at top pressure lately, on the Cooloola issue, the struggle to get national parks in rainforest areas, and many others, as well as on the Reef, and on the production of the magazine. We had organised a week-long school on conservation questions, at Binna Burra – Taff Fenton's special baby, and one that meant much organisation and time. He was also our secretary, and the new interest in the Reef and other questions brought a great deal of correspondence from the public and from our own members, now rapidly increasing in number. Taff had been working particularly hard, and we were now worried about his health too.

It was no fun to be an office-bearer of a small voluntary spare-time organisation like ours, which had taken on so many issues, as well as producing a nationally-distributed magazine. We would have liked to be able to employ a full-time secretary, but the money simply was not there. Poor Taff was in effect doing two full-time jobs, and our Brisbane office was very small.

I was quite unable to do much of my own work, and documents and correspondence piled up, unfiled. But Hilda and Taff Fenton had given over most of their house to stacks of letters, files and copies of the magazine, and every night meant work. Our private lives and our own jobs were suffering. It was no wonder that we were getting weary. No Christmas holidays for us!

We knew, too, that we were up against an almost impossible task in fighting not only international oil companies on the Reef question, but international mining interests too, over the Cooloola mining applications. With our pitiful resources, it was very much a David and Goliath job.

Early in January, John came to Brisbane on his way south. The Littoral Society and Wildlife Preservation Society executives met to talk over tactics, John in a semi-whisper to save his threatened larynx.

We decided that nothing less than a Commonwealth-dominated

authority to administer all the Great Barrier Reef province, plus the moratorium and the scientific survey, would meet the Reef's case. While the Great Barrier Reef Committee now supported the moratorium, it was still leaning towards a policy of 'controlled exploitation' on the Reef. With the May symposium of the ACF now being organised, it was important that any speaker on the Reef's biology should express the views of the young marine ecologists on the danger to the Reef as a whole. We decided to ask the ACF to invite Professor Burdon-Jones, whose views we knew, as a speaker. Would the proposed resolution to be put to the symposium be amended to contain what we had asked for? None of us felt very optimistic.

John went on to Melbourne for his new series of tests. Always cheerful and dynamic, he had put some new heart into us, even in the midst of his personal worries.

It was near the end of January that the big break came, this time in news from overseas. The great off-shore oil leakages at Santa Barbara in California began, making headlines everywhere. The papers were full of awe-stricken accounts of the damage. There were full-page photographs of dead and dying sea-birds, oiled seals and dead fish washed ashore, of blackened beaches, slimy rocks, and volunteer workers struggling to clean up the beaches as more and more oil came ashore on every wave.

Santa Barbara's tragedy was the Reef's good fortune. It made the best possible publicity for the possible fate of those coral reefs and beaches. We no longer had to keep hunting for documents and information, publicising what we could get hold of, emphasising that oil leaks and blowouts really could and did occur, and that the monotonous assurances by the state government and the oil interests were not all they seemed. There it all was, in front-page stories, in articles and photographs.

Now the public began to swing so strongly against allowing similar dangers to the Reef that we felt more than half the battle was won. Letters flowed in to the press, editorials were written, questions asked in Parliament. It seemed that any government with responsibility for the Reef, and for all its values, must at least decide on halting the proposed drilling programmes and commissioning research and an inquiry.

But the state government, at least, seemed immovable. To all questions it simply repeated that there was no danger to the Reef and that 'every precaution' would be taken to prevent any such disaster happening there.

Knowing the Prime Minister, John Gorton's already expressed views, we looked hopefully for some move at the Commonwealth level.

We had good reason to think that much must be going on behind the scenes. But still there was no move, either for a moratorium or on the proposed legislation to give the Commonwealth its needed powers.

And, in spite of our letters, the Australian Conservation Foundation made no comment publicly on the danger to the Reef. Instead, it simply concentrated on organising its May symposium.

Mr Gorton had personally chanced his arm on the side of protection for the Reef. But the old Commonwealth-States confrontation, and the Liberal Party's federalist bent, stood in his way. As far as public opinion went, the Prime Minister would have no problem in taking action, but in other directions he had thorny considerations.

The Santa Barbara leakage had united conservation societies all over Australia on the question of Reef oil drilling. The Littoral Society in Queensland had been hesitant, as had others. But now Des Connell, president of the Littoral Society, told a press interviewer, 'We want to have oil-drilling on the Reef terminated until truly effective means have been found to prevent the release of large oil slicks'. He added that the Society now saw a well blowout as even a greater danger than oil spills.

As for us of the Wildlife Preservation Society, we went further. We simply did not believe that oil-drilling on the Reef was permissible at all, in the light of the Reef's importance as a world possession, and of its ecological unity.

Now the Federal Labor Party declared itself. Dr Rex Patterson, its spokesman for North Queensland, went on record in February. The decision to go on with drilling on the Reef, he said, was 'the height of folly and an incredible act of irresponsibility'.

The State Mines Department's chief engineer, asked in Brisbane by the Select Committee on Wildlife Conservation about what was known of the effects of oil on corals, admitted that he did not know of any research done on the Barrier Reef. He said he would like to see work at the Heron Island research station on the question. Perhaps drums of oil could be poured on the reef there. This did not get a good reaction from the public.

Eighty per cent of the Reef, he testified, was under drilling permits, but the department was taking every precaution to ensure that no blowouts occurred. He did not explain how, in the face of the Santa Barbara experience, this could be done.

There seemed to be an inexplicable reluctance on the part of the Mines Department to go into any details of these precautions, or name the 'experts' whose advice they insisted had been sought.

Answering a question as to what would be done to control an oil spill, if a blowout did occur, he said that dispersal detergents, 'non-toxic to reef and marine life', would be sprayed over the slick area within twenty-four hours. In the light of what had happened to Britain's coast after detergents had been used on the *Torrey Canyon* spill, this seemed hardly helpful. And the detergents used on the Santa Barbara spill did not seem to have done much good either.

When would drilling begin on the permits already issued by Queensland? He answered that an offshore well would be drilled in Repulse Bay, off Mackay, in October next. It would be a combined operation by a Japanese-Australian owned company. The successful tenderers for the other offshore oil-drilling permits would be announced by the Mines Minister at the beginning of March.

The news that drilling was so near startled many people. Now came the first sign that Queensland's Labor Party, too, wanted the drilling stopped. The Opposition Leader called for a ban on oil-drilling and mining on the Great Barrier Reef. 'It is almost criminal,' he said, 'that we know so little of the Reef. The responsibility for preparing the first report on the Reef in almost forty years has been left to a team sponsored by a Japanese newspaper' (the *Yomiuri* expedition).

In the following week, Tenneco Signal spudded in an exploration drill near Darnley Island in the Torres Straits. This was an area under Commonwealth control. And the state government announced that its senior petroleum engineer would go overseas to study the latest oil search precautions. Since most people had innocently believed that these, whatever they were, were already included in the 'stringent' requirements for Reef drilling, this did little to make the public any less nervous.

As for the Tenneco drill, it was obviously going down without the benefits of these latest precautions.

The state Mines Minister, pressed by questions in Parliament and from the press, said that though the names of the successful tenderers for the other Queensland permits would be issued, no actual permits would be given until the authorities were satisfied there was no pollution risk. There was no indication on what kind of evidence would satisfy them. The conditions of the permits would not be decided on, he said, until the engineer had come back from overseas.

The Great Barrier Reef Committee now came in again. 'It's not only the oil wells that are giving us concern,' said Dr Endean. He pointed to the great size of the tankers plying through the Reef's narrow channels. 'Some of these vessels draw about thirty-four feet fully laden. In places the water depth is only thirty-seven feet.' He was also sceptical about the state's capacity to deal with oil-spills on a large scale. Complex and variable currents could carry these in unexpected directions for hundreds of miles. He criticised the state Mining Engineer's expressed views about the effect of oil on corals. would hardly think there is any doubt that oil on coral would kill it.'

Nevertheless, he said, the Great Barrier Reef Committee was not opposed to oil exploration, providing adequate safeguards were taken. What these adequate safeguards were going to be, was a question avoided by everyone, it seemed.

On the Federal front, Dr Patterson now demanded that the Commonwealth Government should give an unqualified assurance that it would not ratify any oil-drilling permit on the Reef. The Minister for National Development answered in soothing terms. The state and Federal governments, he said, were both 'studying closely' two recent problems in offshore oil-drilling (presumably the Santa Barbara and Bass Strait blowouts). The Bureau of Mineral Resources, as well as the Queensland Mines Department, was sending a senior geologist to California. But the question of offshore drilling in Queensland was still 'a matter for consideration between the two governments'.

This clearly indicated that Queensland was taking a hard line to any Federal approaches. Meanwhile the Santa Barbara leaks went on and on, apparently quite beyond control. So did the protest and the publicity.

In an article for *The Australian*, Eddie Hegerl drew the story together. 'To conservationists and marine biologists alike,' he said, 'the issue has become clear-cut – a major oil disaster could not be prevented from destroying large areas of the Barrier Reef.' He pointed out that the Marlin gas blowout in Bass Strait had taken more than a month to rectify. Public concern was very high, and numbers of conservation groups, fishing and skin-diving clubs and service organisations had joined in the campaign. Over 15,000 car stickers had been sold. He added a plea for more support for scientific study of the dynamics of the Reef's flora and fauna, in place of the 'static' studies of the older biology.

The rising public clamour gave the state opposition party a very fertile field to plough. The fate of the Reef was now one of its chief weapons against the government, and it made much mileage.

The split in the Great Barrier Reef Committee between its biologists and geologists grew wider. Dr Orme of the Geology Department of the University of Queensland wrote a letter to the press, saying that proposals to stop oil-drilling were 'sidestepping the issue', since there was already a huge danger from oil-tankers. The implication, that the Reef was already doomed by tankers and that it was therefore unjustifiable to stop the drilling, seemed a queer twist of logic to many. As for us, we knew that the tankers were indeed a most serious danger. But we also knew that they would only be diverted from Reef waters, or provided with better pilotage and other safeguards, if we could raise enough public concern on the question of oil exploitation. Moreover, our overseas informants told us that the length of time of exposure to oil was a critical factor in how marine life responded to it. Therefore long-standing pollution, from constant leaks or spills, would be a worse strain than a big single exposure.

We had plenty of such informants now. Santa Barbara residents, and people who had visited the area, were still in a state of shock and indignation. Our Mackay branch had collected a number of letters from such people, after a letter to the *Denver Times*. They included academics, as well as ordinary people distressed and angered over what had happened, and was still happening, to their beautiful playground. Almost without exception, they urged Australians not to let the same thing happen to the Barrier Reef. They had started a strong citizens' campaign, under the slogan Get Oil Out, or GOO for short; but it was a hopeless task, now that the industry was actually installed and the oil being used. *Keep* oil out, instead, they told us.

At last we had a letter from John in Melbourne. The tests had shown that his throat condition was almost completely cleared up. 'I feel like The Man They Could Not Hang,' he wrote; he was setting to work again on the Reef question. Sir John Barry, who had coined the useful phrase that if things went on as they were, Australia would soon be 'a quarry surrounded by an oil slick', wanted to see him, and he was to give evidence to the Senate Select Committee on Offshore Petroleum Resources as soon as he could. The reprieve had given him 'an enormous increase in useful energy'.

We were thankful; we needed that energy. Apart from all our other work, we had been besieged by correspondence from many people concerned for Reef welfare and had had to deal with a great number of new applications for membership to the society. We were no longer seen as sinister enemies of Progress and Development, or agents for overseas interests, by most people, but there was still a great deal to be done. Having John and his fighting spirit back in the field was, we all felt, nearly essential.

5

Geologists versus Biologists

The Japanese 'submersible' vessel, *Yomiuri*, set out on its research voyage in January. The amount of involvement by Commonwealth and state government departments interested in mining and drilling was far from reassuring to those of us who were concerned with the protection of the Reef and its waters. There were thirteen scientists from the Commonwealth Bureau of Mineral Resources, working on the geology of the continental shelf in Torres Strait, the Great Barrier Reef and the Arafura Sea. Eleven other scientists were working on separate research projects, including the Crown of Thorns plague.

The Crown of Thorns plague was still a matter of public and scientific controversy. Endean's recommendations on banning the collection of triton shells had not yet been put into effect by the Queensland government; but some other biologists felt that these were, though probably useful, not going to do much either to control the plague or solve the question of what was causing it.

As far as we could find out, the biologists' research programmes would not have much bearing on the biological dynamics of the Reef – indeed, they scarcely could, in the short time of the survey. It would certainly not come anywhere near fulfilling the need for the thorough scientific survey we were asking for; but we were afraid it might be accepted as doing so.

Still, by now we had all kinds of organisations joining in the campaign for the Reef – museums, universities, conservation groups, even fishermen. Most of them, too, were more interested in the biological and aesthetic importance of the reef than in its geology. The continuing Santa Barbara leaks were still in the news; so was the indignation of Californian citizens.

We went on collecting information on the Santa Barbara situation both from US publications and from Santa Barbara's residents. They, too, had been told that oil spills 'couldn't happen'; they were outraged that their protests and requests for stopping the drilling, before the leaks broke out, had been ignored.

During 1968, one of the earliest of the drilling rigs of offshore Queensland, the *E.W. Thornton*, had worked on the Swain Reefs, striking much bad weather. (Little had been heard of this project, and we had not found out anything about it while it was still in progress.) Louis Salzman was an engineer who had worked for five months on this rig; he now wrote a letter to the press from Santa Barbara, where he had witnessed the disaster at first hand. He had just been reading the evidence given by the Queensland state Mining Engineer, Mr Morley, to the Senate Select Committee. Mr Morley had said that the Santa Barbara blowout had not affected marine life; and he had emphasised the safety precautions taken by Australian offshore drillers. Salzman said in his letter:

Time after time, bad weather caused breakdowns that would have caused oil spills if we had struck oil. As it was, thousands of gallons of drilling mud were released into the water, and all kinds of scrap was dropped overboard during the drilling operation. [He enclosed with his letter many photographs of oil-clogged beaches, dying seabirds and oil-affected marine life.]

As the Santa Barbara area deteriorates, marine life dies, and birds struggle for life, I keep thinking that the same thing could have happened while we were working on the *E.W. Thornton*. Perhaps Mr Morley should visit California and see for himself. It is not only the coral that might die, but the beaches turned black, the white birds turned black, boats covered with oil, the smell, and the dead marine life. How many turtles could exist if the surface of the water was covered with oil? What effect would oil-covered beaches have on the turtle eggs laid each year?

Australia has something that exists nowhere else on the face of the earth. The idea that anyone would take the remotest chance of damaging the Reef is absolutely beyond belief.

There seemed to have been no controls over waste disposal, and we wondered whether there were any controls on drilling methods either. Meanwhile, the political and industrial battle behind the scenes kept things outwardly quiet for a while. We kept up the pressure as best we could, with letters to press and politicians. John Büsst was preparing his submissions to the Senate Committee on Offshore Petroleum Resources, with little time to spare. And our secretary, Taffy Fenton, was now ill and unable to carry on with the burdens of the work he had been doing for us. He was a devoted battler; we felt his loss, and had a job to find anyone to take his place.

There was a state election coming up in May, and the state government had begun to realise the extent of opposition to drilling and mining the Reef. They announced towards the end of March that 'no new prospecting permits' would be issued for the Great Barrier Reef area until further knowledge of possible damage was obtained a promise that seemed to have little bearing when the extent of present permits was considered, but was a useful election move. They had never announced the names of the successful tenderers. They were apparently waiting for the return of their mining engineer from abroad, which was scheduled for June or July.

The press, cautiously welcoming this, took the opportunity to urge that until there was a foolproof system of drilling, the government had better forget about drilling in the Great Barrier Reef area; but the government made no answer.

The Minister for Mines had said in Mackay that he had been told by a 'world authority' that the risk of an offshore blowout was 'about the same as a meteorite hitting a city'. He had said, too, that the Santa Barbara blowout was the only one to occur, though thirty per cent of oil supplies came from offshore drilling. We countered with the actual figures on this. There had now been 100 oil incidents in Cook Inlet, Alaska, alone, and numerous others elsewhere. The Red Adair trouble-shooting team had been called to eleven blowouts offshore during the past year, and the Santa Barbara leaks were numerous in themselves.

John recounted all this in his submission; and he pointed to the frequent cyclones that struck the Great Barrier Reef. 'It is not a pleasant thought to contemplate the destruction that could be caused by, say, ten drilling rigs on the reef, caught by two successive cyclones, as may often happen in our monsoon season.' He again used the analogy between mining the Reef and breaking up the Taj Mahal for road metal (as he had done in his earlier letter to *The Australian*); and put our case for complete conservation measures for the whole of the Reef, under Commonwealth control, as the younger marine biologists had given it to us. It was useful to be able to get all this information lodged with a Commonwealth investigating committee.

Early in April the state government Opposition revealed that an American oil search company was carrying out an aerial survey between Mackay and Cairns, with eighteen men and a DC3 aircraft. They asked the government to give all details of the operation, and name the companies for whom the report was being made. Clearly the oil companies were not much discouraged by the situation.

But with the government in the firing line before the May elections, the Mines Minister answered simply that an authority to prospect was not a lease to drill, and that if applications to drill were to follow the prospecting 'they would have to have a very good case'. Indeed; the government kept on emphasising that they intended to see the Reef was not endangered.

The fact that most of the existing permits had only recently been renewed, under the new Uniform Offshore Drilling Act, and that one company was already standing by waiting for equipment from overseas to arrive, scarcely seemed to support this. But there were still people who believed that the government would finally refuse drilling applications, now that the dangers had been so effectively illustrated by Santa Barbara.

And there were also plenty of influential scientists who were in favour of Reef drilling – they were not biologists, but geologists. They now came into the argument with remarkable simultaneity. One Rhodes Fairbridge, professor of geology at Columbia, New York, came to Australia from a study of Torres Strait reefs, and gave a lecture to the Royal Society in Brisbane. He was reported as saying that the Reef should be exploited 'immediately and to the hilt'. Those who wanted it protected were no more than 'sentimentalists who want to put a barbed-wire fence around everything'. Ninety-nine per cent of coral reefs, he asserted, were 'dead anyway'. A biological survey would take ten years or more. If anyone was still worried, oil companies could be faced with heavy fines for pollution.

The Queensland government, when asked for comment, merely reiterated its promises of Reef protection, though never committing itself to how this was to be done.

Now a British oil geologist, a lecturer at Queensland University, told the Senate Select Committee that he agreed with Fairbridge that the pollution risk from drilling was 'insignificant'; the first round of drilling probably wouldn't involve more than ten or twelve wells, and 'the more promising the results, the more willingly will money be spent to ensure that production facilities do not spoil the beauty and amenities of the Reef'. The time to drill was now.

Dr Chapman admitted that he had only been in Australia a few months and had seen the Reef only from the air. No, he knew of no research being conducted anywhere on the effects of oil on coral.

All this kept the controversy going. Dr Frank Talbot, himself a marine biologist of note and a member of the executive of the Great Barrier Reef Committee, pointed out in a press interview that most people were interested in the Reef as living material, not as a 'dead structure'. He had 'nearly burst a blood-vessel' when he read of Fairbridge's views. He emphasised the lack of biological research on the Reef, and the need for a thorough survey.

The Reef had turned into as complex a partisan battlefield as it was a complex ecosystem. Commercialisation versus protection, geologists versus biologists, even biologists versus biologists, and Commonwealth versus states; the arguments and counter-arguments raged on. But the people who might be said to own the Reef – if anyone did – were almost unanimously against the drilling. Their letters to the papers asked why there was all the hurry to drill? Surely if there was any danger it was better at least to wait and find out what could safely be done, and if nothing could, then the Reef ought to be left alone. The oil would not go away, and better drilling methods might be devised, if drilling finally had to be done.

The next major operation by conservationists was to be The Australian Conservation Foundation's symposium at the beginning of May. We had originally welcomed this idea enthusiastically, but now we were beginning to have our doubts.

The object of the symposium was supposed to be to provide a forum for all the varying views. In fact, as we knew privately, it was intended to lead up to the resolution recommending a joint Commonwealth-state commission, which would only be advisory, to examine and assess development proposals, research needs, and planning principles. And the question of who really owned the Reef, or had authority to issue or refuse mining and drilling permits or protect its living organisms, was to be crucial. But we did not like the look of the programme.

It was to be a one-day symposium, with representatives from the Commonwealth and state governments, all of whom were to speak; the other papers were to be given by a professor of geology from Sydney

University; an elder biologist from Queensland, an influential member of the Great Barrier Reef Committee, who we knew still believed the state government's protection promises; the exploration manager of Conzinc Riotinto Exploration (another geologist, we noted); the manager of government relations for Esso Standard Oil, and the general manager for TAA.

Frank Talbot was to sum up. But there was no speaker with connections with North Queensland (though we had asked for Professor Burdon Jones to speak), and no marine ecologist; we feared that there would be no one who would press for full protection of the Reef. And it was a very heavy programme, with no provision for an open session of discussion. We were not in full agreement with the draft resolution and we wanted important amendments. But would there be time or opportunity to get those amendments through?

We were pretty sure that John, Len and I, as well as the Littoral Society, were not wholly popular with the top echelons of the ACF. In spite of the unanimous resolution passed the year before, which should have been official ACF policy, there was no mention, in the resolution as drafted, of the need for a moratorium and for a thorough scientific survey. I wrote to the Director, Dick Piesse, expressing our disquiet at the way the symposium seemed to be shaping; so did John. The reply indicated that nothing would be changed at this stage.

Vin Serventy wrote me from Sydney: 'Basically the ACF has fallen into the trap I thought it would. Once you are dependent almost entirely on Government money you are in the doldrums. It would have been better to have had to live from hand to mouth ... than to accept federal aid. One doesn't bite the hand that feeds.'

We were not worried by the possible Commonwealth government attitude; by now we were fairly sure that that would be favourable to our case. But to have all those speakers on the side of Reef exploitation seemed to us to be asking for the kind of publicity we were afraid of.

I wrote to Francis Ratcliffe, answering a letter he had sent me: 'I agree with all you say about the importance of getting a statutory body set up; not only as a "watchdog", but with some real authority. Otherwise, I am quite convinced that it is only a question of time before oil-drilling permits will be granted.' I pointed out, too, that North Queensland conservationists, who had been so active in getting the whole question of the Reef into the limelight, were not represented at all and were consequently sore; particularly as they had expended so much time and money, would have to attend from a great distance if they wanted to put their views, and were to be faced with at least two, and more probably three, advocates for Reef exploitation on the speakers' list. The fact that there was no open session was especially worrying and the time allowed for questions was much too brief. All our northern branches were in a state of indignation, and I was in a specially awkward position, as President of the Society, and a Councillor of the Foundation.

I decided I would not go down. There would be plenty of Queensland representation of the society, headed by John; and of the Littoral Society too – they were more than able to put our views, and if I stayed away it would underline the society's objection to the programme and the proposed resolution. All of us had to pay our own way to Sydney, and money was scarce and better kept for the work we were doing.

Moreover, we were pretty sure that southern conservationists were taking the same view of things as we were. The need for preventing oil-drilling, and the Santa Barbara disaster, had brought most of them into the battle on our side; and the ACF's programme was now seen as much too cautious.

John had worked out the wording of our proposed amendment to the resolution, and took counsel with the Littoral Society members who were going to the symposium – some of them highly qualified young biologists. The Queensland contingent set out, determined at least to let the Sydney press, and the rest of Australia, know that there were people who opposed the exploitation of the Reef, and could answer the arguments of geologists and miners with better ones.

We were now known as 'vocal' and 'obdurate protectionists'; we wore the label with pride. We had had enough letters, from people overseas as well as in Australia, supporting our stand – let alone the results of the 1968 petition – to know that most people wanted the Reef protected.

The Commonwealth spokesman, Mr Malcolm Fraser, said that the government regarded the Reef as a priceless asset, and it would 'use all the powers it had to prevent the Reef from being despoiled'. The state government spokesman said in reply that the Queensland government had already approached the Commonwealth with a view to establishing a Great Barrier Reef Resources Advisory Committee and was awaiting the reply 'with patient anxiety'. (Of this, more was to come later.)

Professor Maxwell, after outlining the Reef's geology and hydrology, ended with the emphases we had expected. 'The possibility of excessive reef destruction by animals – human and non-human,' he said, 'would appear to be quite remote.' The enormous dimensions and viability of the Reef province 'should not be obscured in the hysteria of ignorance', the Reef represented a 'major natural asset' that invited thorough appraisal and responsible exploitation.

The biologist, Professor Stephenson, emphasised both the immense wealth and variety of species, and the complexity of their inter-relationship. There were two main problems, even in the small amount of biological work already done, he said: the collections made on the Yonge expedition, which had been the most important and had been undertaken in the 1920s, had not even yet all been identified; and they were far from comprehensive. There was very little support for taxonomic work on such collections. Knowledge of diversity on the Reef was woefully incomplete; but what had been done indicated the complexity − 400 fish species had been recorded from Heron Island alone. He pointed out that pollution was a great danger to such highly diverse communities − 'it seems almost certain that the more diverse the biota, the more sensitive the situation will be to pollution. The more diverse situations in the marine world, the Barrier Reefs are likely to be proportionally the worst hit'. He referred to limestone mining, commercial fishing, angling and spearfishing as likely to cause upsets in the Reef's communities, and 'oil-drilling carries obvious dangers of a wide-spread catastrophic effect'. He concluded that basic biological research and a 'sane and justifiable programme of conservation' were required.

The trouble was, as Dr Grassle pointed out in discussions, that there were two unanswered questions of crucial importance to the whole discussion: how far was the Reef an inter-related system, where interference to a part might disturb the whole; and how irreversible would the effects of disturbance be?

The research needed to establish either fact one way or the other might take many years, much biological manpower, and far more money than was likely to be allocated to a scientific project that might come up with answers embarrassing to enterprises that wanted to use the Reef for their own purposes. We were, in effect, in a double bind; without that research it was clear that the Reef ought not to be risked, but the research was never likely to be made definitively unless it was supported by quantities of money that were unlikely to be made available by governments and enterprises dedicated to the kind of projects and uses that the research might prohibit. The fact that geological research was so advanced, biological research so starved, was one result of this.

Sir Percy Spender, who had just come back from overseas, where he had been President of the International Court of Justice, gave the talk that was to be the day's highlight. His view was that the boundaries of the Australian states ended at low-water mark. 'Consequently the states have no rights over either territorial seas or the mineral and other natural resources of the seabed.'

This was to be the beginning of a legal argument, and a battle between states and Commonwealth, which was not to be decided for many years to come. But he suggested that the issue should be determined 'with despatch', by the Commonwealth's instituting proceedings against a state, for example Queensland, before the High Court of Australia.

The other speeches were predictable. The Commonwealth Minister for Primary Industries gave an account of the Commonwealth's rights and responsibilities for fisheries in Reef waters and of the protection it would be able to give under the Continental Shelf (Living Natural Resources) Act, so recently passed. The spokesman for limestone mining said that 'it should be possible to carry out mining operations on restricted parts of the Reef without interference to actively growing parts of the Reef'. The spokesman for Esso Standard Oil said that, though no one could guarantee that blowouts or oil spills would not occur, there was a new non-toxic dispersant, Corexit, that was hopeful.

The informed part of the audience grew more and more restive and noisy. Question time was predictably far too short; speakers had overrun their time, and there were many things in the talks we had no time to challenge.

Dr Talbot gave the summary, emphasising the need to decide now what was to be done with the Reef. Mining and oil extraction were short-term operations with dwindling assets; fishing and tourism could be continuing and growing assets. He quoted Dr Fred Grassle – who was in the audience as a member of the Queensland contingent. Could we use different sections of the Reef for different activities? 'The Great Barrier Reef is rich in large measure because it is large, and if it is sliced into segments this would drastically reduce its richness. We therefore have to think of the reef in its totality.'

And he pointed out that the science of ecology was in its infancy. 'We do not know for certain what will happen to a stream with three fish species if we add a fourth – how can we guess at what would happen to the Great Barrier Reef if, for example, we introduce a slight but continuous pollutant?'

The formal resolution was handed out, and the Chairman asked for 'comments'. But the meeting, he said, had to end at 6 pm, and it was now very late.

John Büsst and Vincent Serventy put forward their amendments. No mining should be permitted for at least ten years, or until such time as the advisory body declared that it was completely satisfied that no damage would be done to the Reef by mining for oil, limestone or other minerals. The Chairman refused to accept amendments to the resolution. Dr Grassle, trying to move another amendment which would have further increased protection for the Reef, was unable to do so.

By now the audience was in something of an uproar. But the resolution was put forward as it stood and no further discussion allowed. With the resolution passed, the meeting was closed; but it did not accept this calmly.

The returning Queenslanders were indignant. John described the symposium as a 'bloody shambles'. No speaker had been allowed to put the case against oil-drilling and for a moratorium; the resolution had been pushed through, it seemed to us, unjustifiably; Grassle had not been allowed his say.

The Littoral Society wrote a letter of protest to the Foundation; the Innisfail branch, writing in support of it, said:

> [The last meeting] fully supports the letter forwarded to you by the Queensland Littoral Society, concerning the recent Symposium on the Future of the Great Barrier Reef, and regrets that the Symposium permitted neither time nor opportunity for discussion on extremely controversial points raised by geologists and others in favour of the commercial exploitation of the Great Barrier Reef. Further, that time did not permit Mr J. H. Büsst's motion to be voted on, nor was Dr J. F. Grassle's motion even allowed to be put to the meeting.
>
> In view of this and the inadequate provisions made for these extremely important discussions from the floor of the meeting, we feel that the resolution as passed was not representative of the opinion of the majority of those attending the symposium, and we wish to place these matters on record to be included in the final printed summary of the Symposium. [It was not.]

We in turn supported this. Partly as a result, and partly because there had been so much obvious and vocal feeling expressed at the meeting on the subject of drilling and mining – not only we, but plenty of people attending it from Sydney and Melbourne, had put this point across pretty forcibly – the letter the Foundation actually sent to the Prime Minister did in fact convey the substance of the amendment John had moved, and the strength of the feeling for protection of the Reef against all such dangers.

So we had, in effect, won another point. But the symposium had done damage, as well as good. We did not doubt that it had been planned, as the ACF then saw it, with a good purpose, to indicate to its sponsors and financers that conservationists were willing to listen to the other side and that the ACF was willing to be a mediator in disputes of this kind and adopt an 'even-handed' approach; we personally respected this view. But it had resulted in much publicity for claims that were demonstrably inaccurate, such as the view that protectionists were ignorant and hysterical, that the Reef could safely be mined and drilled with 'certain precautions', and that damage could be restricted to certain areas. The arguments of the ecologists had scarcely been touched on; but they were vital ones and needed publicity.

The main damage, though, was to the Foundation itself, in the eyes of many who were convinced that the Santa Barbara disaster meant that the Reef must be protected from any such possibility. The address of Professor Maxwell, in particular, with its implication that responsible scientists looked on protectionists as foolish, had upset not only conservationists but those in the Commonwealth Government who were listening to public opinion and were willing to take responsibility for the Reef. The way the programme had been arranged, with its crowded addresses and lack of open discussion, had antagonised many even of the Foundation's own members.

Above all, a gap was showing between the conservation societies and the ACF, which purported to be their national spokesman and leader. If this was leadership, it was from behind. We had lost faith in the Foundation; and it showed very soon in a notable falling-off in the drive the Foundation was making for new members. Probably it resulted in increased membership by industrial and mining interests which now saw the Foundation as, if not an ally, then a useful forum. That was to damage the real direction and influence of the Foundation for years ahead, and in other fields than that of the Reef battle.

But now we were back in Queensland, and the May election speeches were going on. 'Protection' of the Reef was the government's theme.

Fred Grassle gave evidence to the Select Committee on Offshore Petroleum Resources. He called for a halt to all oil-drilling in the Reef area. The Reef, he said, was the most complex, varied and productive biological system in the world, and was of 'incalculable value'. Areas of the Reef affected by oil might 'never recover and would be very unlikely to recover in our lifetime', and even if there turned out to be extensive oil fields there it would still not make good sense to sacrifice a resource such as the Reef, whose importance would grow with each generation, for a short-term profit from a short-lived industry. Tanker spills and pollution from the mainland were also serious dangers.

My own and other work on the Barrier Reef indicates that the diversity of species on the Reef is much higher than in any other aquatic environment. This means that where in other environments we find hundreds of species, on the Reef we find many thousands. For example, the reefs of the Caribbean have only about 40 species of coral whereas the Barrier Reef has at least 200 (the actual number must await the results of the first collection of coral from many parts of the Reef made recently). In my own work, I found over 200 species of polychaete worms in a single 13 pound lump of coral – this is more than the number collected in extensive surveys over long periods of time in other regions. Every instance of marine pollution has resulted in a reduction in the diversity of plants and animals. The Great Barrier Reef is more susceptible to pollution than other environments because even the slightest changes in the environment can result in tremendous reduction of diversity. The reason for this is that all the species of the Reef are highly specialised in their requirements. This specialisation results in many complex interactions of plants and animals with each other and the environment. We can think of these inter-relationships as many fine webs of mutual dependence which are distributed and stacked in intricate spatial configurations. Any slight disturbance of such a system would have tremendous consequences.

It has been said that parts of the Reef might be separated and protected as self-sustaining systems. It is unlikely that any part of the Reef can maintain itself in the same state if separated from the whole. Recently, Scheltema (1968) found that larvae of zoanthid corals are transported across the Atlantic. The existence of these long-distance larvae raises the possibility that the reefs of the Pacific are part of a large system in which the Great Barrier Reef plays the dominant role. Preston (1960) has shown that the number of species found in large isolated areas such as islands and continents is directly proportional to the size of the area. This may be one reason the reefs of the Pacific have so many more species than the reefs of the Caribbean. If the area of the Barrier Reef were reduced, then the whole Reef and even many of the Pacific atolls might be affected.

After examining what was known about the effects of oil pollution, and sharply criticising the use of detergents to break up oil slicks, he pointed out the urgency of research programmes and suggested that:

All oil–drilling on the Great Barrier Reef should be stopped until it can be guaranteed that oil pollution will not occur. The present drilling constitutes a threat and if oil is discovered, the risk will be much greater both from wells and the tankers which service the wells. Strict control of tanker shipping in the region of the Great Barrier Reef must be maintained. Since we know so little about the Reef and the effects of man's interference, a commission or other Commonwealth body should be set up to decide on all matters related to the Reef. Some of the important characteristics of such a commission should be:

1 that it assume full responsibility for all forms of pollution and damage to the Great Barrier Reef.
2 that it have a large research staff competent to examine all consequences of pollution including research into the functioning of the Reef.
3 that it have sufficient funds to award grants to individual scientists or organisations for specific research proposals.
4 that it consult with, but be independent of, all existing government departments.
5 that it should have the power to recommend legislation and the means for seeing that laws related to the Reef are observed.

The next day, the state Opposition leader pointed out that an American company had actually been drilling on the Reef over the last few weeks, in spite of the government's assurances of protection. The drilling rig, the *Glomar Conception*, had sunk its well to 12,000 feet in the far north of the Reef, at the Darnley Islands in Torres Strait. The Opposition, he said, would withdraw every authority, lease, and drilling permit if it became the government, pending a full scientific inquiry.

Another biologist, Howard Choat, now wrote to the press pointing out that proposals for a few marine national parks, with drilling and mining allowed outside them, were fallacious. The plankton, which was largely composed of eggs and larvae of most of the Reef's plants and animals, moved on the surface currents, which were complex and little studied. The patterns of plankton movement were still 'completely unknown'. Thus a few separate marine parks would not at all necessarily be self-sustaining entities; they were connected inseparably to areas outside them, in still unknown ways. Such parks would 'contribute nothing to the main problem – how do we maintain the Great Barrier Reef as a viable biological unit?'

The question of the Premier's big oil holdings, and of the application of Exoil No Liability – a company in which he had large interests – for an important lease on the Barrier Reef itself, was one of the Opposition's main points of attack. The Queensland Premier long ago had made it clear that he was not going to interfere with the 'rights' of his Ministers and himself to take shares in any company. It had been a point of dispute ever since.

The Premier replied in a press election advertisement: 'Is the search for oil on the Barrier Reef a deep election issue? I think not.' He quoted opinions expressed by Dr Endean, Professor Fairbridge and Dr Chapman, and went on: 'The whole matter of oil and the Reef is totally controlled by Commonwealth-states uniform legislation. My government has given an absolute undertaking that we will conserve the wonders of the Reef ... If Queensland were to become self-sufficient in oil, our development and prosperity would be fantastic. Think of the benefit to both our national economy and defence security.' And he declared that if the contracts for oil exploration were repudiated, the state 'could become liable to compensation payments of many millions of dollars. The government's integrity on the matter of conservation and development of the Reef is beyond question.'

There were plenty of people to doubt the Premier's assurance that, in the case of the Barrier Reef, conservation and 'development' of the kind he apparently had in mind were compatible. And his assurances that 'the most rigid conditions' would apply to exploratory drilling seemed to have already been contradicted by the report of what had happened on the Swain Reef drillings. The Reef was, indeed, an election issue; but it did not carry the day. The government was re-elected.

A few days later the Premier announced that the combined Ampol-Japex test drilling in Repulse Bay, north of Mackay, would go ahead in October.

There was a furore. The Premier simply continued to assure everyone that 'the most stringent precautions' were being imposed on the drilling. The chairman of Exoil, C.W. Siller, in an interview with the press, said that 'if an accident did occur it would not destroy the whole of the Reef'. If there was oil 'let's develop it for the people of Queensland now'. He revealed that Exoil was completing a seismic survey on its Princess Charlotte Bay permit area, and 'if this proves positive we will go on from there'.

The marine biologists of the Great Barrier Reef Committee, who had believed the Premier's assurances that though the exploration permits could not be repudiated, anyone wanting to convert them into oil lease would have to have a 'very good case', and that this meant that no oil leases would be issued without the closest examination, were disillusioned and distressed. Apparently Mr Siller would not have to make a case at Nor, presumably, had Ampol-Japex. If the oil was there, it would be developed – that was all.

The heat generated by the announcement was high indeed. All over Australia, newspapers, radio and television ran articles and comment on the proposal. Vincent Serventy in Sydney published an article recalling the statement made by the engineer who had worked on the Swain Reef drillings. Siller, he went on, had claimed that the new 'non-toxic' dispersant, Corexit, would deal with any spills without damage to marine life; but there was no organisation to deal with oil spills and Corexit itself was now under suspicion. It had been withdrawn from use at Santa Barbara.

Serventy quoted a Federal government statement that though various government departments had been examining the possible use of this chemical, 'no firm decisions as to the most suitable detergents have yet been arrived at'. This meant that in eighteen months since the detergent had arrived on the local scene, nothing had been done that proved its suitability.

Since there was no emergency organisation to deal with oil spills, particularly outside ports, it could be days before the detergent even arrived at the disaster area. 'In such an event,' Serventy wrote, 'a lapse of several days could mean a million muttonbirds dead, or a hundred miles of coral reef.' And he pointed out that 'when a small quantity of oil was spilled into Botany Bay in 1967, despite immediate action by a ship's crew spraying the spill with a detergent, the oil could not be contained and was washed ashore along the whole of Kurnell Beach, along Sans Souci and Brighton Beaches, in mangrove swamps and up the George's River as far as Taren Point Bridge. Cleaning up went on for a whole month and 200 men were employed. All this for a tiny oil spill of less than 500 tons'.[5]

If the Queensland government had hoped that their election win would allow them to go ahead unhindered with oil-drilling programmes, they were quickly corrected. We set to work with more letters to the press, demanding Commonwealth action. As we had feared, the ACF symposium had resulted in plenty of publicity for the statements made by geologists and industry representatives. This had to be corrected as soon as possible, but we were now working more than ever against time.

[5] *Sunday Telegraph*, 4 May 1969.

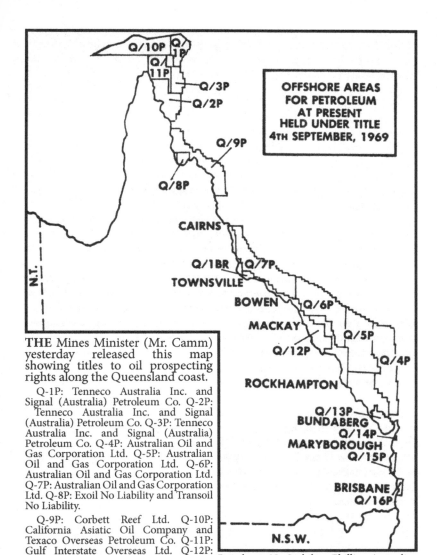

OFFSHORE AREAS
FOR PETROLEUM
AT PRESENT
HELD UNDER TITLE
4TH SEPTEMBER, 1969

THE Mines Minister (Mr. Camm) yesterday released this map showing titles to oil prospecting rights along the Queensland coast.

Q-1P: Tenneco Australia Inc. and Signal (Australia) Petroleum Co. Q-2P: Tenneco Australia Inc. and Signal (Australia) Petroleum Co. Q-3P: Tenneco Australia Inc. and Signal (Australia) Petroleum Co. Q-4P: Australian Oil and Gas Corporation Ltd. Q-5P: Australian Oil and Gas Corporation Ltd. Q-6P: Australian Oil and Gas Corporation Ltd. Q-7P: Australian Oil and Gas Corporation Ltd. Q-8P: Exoil No Liability and Transoil No Liability.

Q-9P: Corbett Reef Ltd. Q-10P: California Asiatic Oil Company and Texaco Overseas Petroleum Co. Q-11P: Gulf Interstate Overseas Ltd. Q-12P: Ampol Exploration (Qland) Pty. Ltd. Q-13P: Shell Development (Australia) Pty. Ltd. and Pacific American Oil. Q-14P: Shell Development (Australia) Pty. Ltd., and Pacific American Oil Co.

Q-15P: Amalgamated Petroleum No Liability, Phillips Australia Oil Co., and Sunray DX Oil Co. Q-16P: Amalgamated Petroleum No Liability, Phillips Australian Oil Co., and Sunray DX Oil Co.

Permits Nos. 1-8 inclusive are for six years from September 1, 1968; Nos. 9, 10, 11 are for six years from October 1, 1968; No. 12 for six years from December 1, 1968; Nos. 13, 14 for six years from January 1, 1969; and Nos. 15, 16 for six years from April 1, 1969.

Oil lease areas (by courtesy of the *Courier-Mail*, Brisbane).

71

With the Littoral Society, we wrote to the Commonwealth Government, and to the press, appealing for action in terms of the actual recommendations of the Foundation, and 'in particular, the Foundation's request to impose a moratorium on drilling and mining'. We said that 'any result of the damage would no longer be laid wholly to the account of the state government in the public mind, should the Commonwealth fail to take the action so strongly urged upon it'. And we drew attention to the deep concern, not only of Australians but of overseas conservation bodies (with which we had been in correspondence) that such action should be 'immediate and effective'. We used radio and television appearances as much as we could. Our time seemed now to be completely taken up with the battle for the Reef.

But we were hoping for the best. Both Malcolm Fraser's statements when he had opened the symposium, and the Spender address on the legal position, seemed to indicate that the Commonwealth was both willing to move and also had the power to do so.

Still, we wrote once again to the Premier of Queensland, appealing for joint Commonwealth-state action to declare a moratorium to allow for a thorough programme of scientific research, and pointing out that this would also allow for further developments in drilling techniques, (now admittedly so unsafe). 'We are entirely unconvinced,' we wrote, 'that the present controls, either by state or Commonwealth, are sufficient to safeguard the Reef, or that the penalties at present in operation are sufficient to repair damage, even if any way were known of doing this.'

As was becoming usual, the Premier made no reply. We had expected nothing; we got nothing.

The *Yomiuri* had now returned from its voyage reporting 'terrible damage' to the Reef by the Crown of Thorns starfish. All the way from Palm Island in the north to Broadhurst Reef in the south, the scientists reported, there was extensive damage; some reefs appeared 'completely dead'.

A few people began to think that if the Reef was as badly off as this, oil-drilling could make little difference. Others argued that this was a side-track – that the Crown of Thorns was a 'natural phenomenon' and so the Reef could be expected to recover from it as it might not from pollution. The question whether pollution itself had set off the plague, by affecting the predators of the starfish and allowing it to increase unchecked, was brought up again. There was plenty of such pollution already in Reef waters; Queensland streams and rivers were loaded with silt, mining wastes, cane-mill effluent, fertilisers and pesticides from farms. (The *Yomiuri* report was later corroborated by one of the government's own marine biologists, who had spent the time from May 1966 to August 1969 on a research programme. He estimated that more than eight per cent of the Reef was then affected by the plague.)

We had no satisfactory reply from the Commonwealth Government to our pleas for intervention. But there was a Federal election coming up in October. We wanted both the present government, and the Opposition, to take a firm stand on the oil question; there was to be no let-up on the pressure.

It was pressure on us, as well, and on our scarce resources; but we were gaining wide sympathy, much more Society membership, and enough donations to survive the financial problems. And all our branches, all along the coast, were hard at work too, as were new branches of the Littoral Society in northern Queensland. We had plenty of hands and minds on the job.

And almost every overseas visitor, canvassed for an opinion, was against the Reef drilling. A visiting American, Professor Wilbur Jacobs of the University of California, told the Townsville University College that the reassuring statements made by the oil industry and the state government about the lack of danger in drilling and the benefits to come seemed 'very familiar'. 'We had the same assurances from the oil industry and reassurances from the Department of the Interior that there was absolutely no danger of pollution to our beaches. Now, scientists are estimating that it would take fifteen years for the area's ecology to get back to normal – if there was no more pollution. But, more than 200 days after the first disaster, oil is still leaking.'

The press was now demanding that the state government clarify its position about the Repulse Bay drilling. The site was a few miles offshore, said a Brisbane paper's editorial, near the gateway to the Whitsunday Passage, in the heart of tourist territory, and one of the great unspoiled regions of beauty in the world.

The drillers were perfectly confident, it seemed. The rig was ordered, the plans were set for October. And the Commonwealth Government, in spite of Mr Fraser's speech at the ACF symposium a couple of months before, was silent.

6

People Take a Hand

August was the time of the Brisbane Exhibition, when we ran a display for the Exhibition Week. Selling the Save The Reef stickers this year was a fast and furious job. Our membership rose quickly. Concerned people stayed at the display for long discussions on the oil-drilling question. Some went down to the State Tourist Bureau display, not far off, and plastered the showcase of corals with stickers, which were taken off as fast as they were put on.

The State Liberal Party was to hold its annual conference from 22 to 24 August. The agenda looked a lively one for the Reef question. Eight motions on conservation were being brought up, and most of them referred to the Reef. One asked for a moratorium and biological survey, one for the convention to support the ACF's suggestion of a joint state-Commonwealth advisory board.

The Prime Minister was to open the meeting, and he had already stated his personal views. We hoped he might influence the convention. We knew that some state Liberal Party backbenchers were anxious and upset about the increasing criticism of the coalition government. The Country Party, to which both the Premier and the Mines Minister belonged, was unlikely to change its own policy.

The Liberals could hardly split the coalition without bringing down the government in the process. And after all, the coalition had only just been returned at the polls. It did not look as though the convention would be much more than a forum for the discontented.

But on 16 August, a week before the convention, the backbenchers did succeed in forcing a joint meeting of the government parties to discuss the Reef drilling. The cabinet argued that the government could be up for 'hundreds of millions of dollars' if the oil leases were repudiated now. The meeting ended without a vote being taken, and with apparent victory for the Premier.

Two days later, however, the Mines Minister made one of his unfortunate public statements, which only succeeded in getting the government further into the mire. He said that, while he was confident in the precautions being taken, no oil spill could in any case reach the Reef, since the currents would carry it northward, not eastward, from Repulse Bay.

This set off a fresh outcry in the papers. The *Courier-Mail* pointed out that if oil did 'spew from an offshore well blowout, it would not be the Barrier Reef but the fringing reefs of tourist islands such as Lindeman and Brampton, sixteen miles away, which would be damaged'. Nobody, said the writer, would want to come to an island where their swimming, fishing and sunbathing were likely to be spoiled by black oil-slicks. Anyway, more than one drill-hole was involved; and the Japex drilling would set a precedent. How would the government, once one drill had gone down, resist the pressure for more?

'Everything,' Mr Camm had said, 'involves a risk' – driving a car, clearing land for a farm, or drilling for oil. No, he did not know of a foolproof method of offshore drilling. But the Mines Department, he revealed, would have an inspector on the rig twenty-four hours a day.

The Opposition spokesman on conservation asked what the inspector would do in the case of a blowout – put his finger down the drill-hole?

The Mines Minister now argued that to repudiate the agreements would cause 'worldwide loss of confidence in Queensland's integrity'; there would be 'a flight of capital from the state', while an important oil-field on the Reef would mean 'a great increase in investment, employment, business activity and government revenue'. This was unsupported by any economic arguments, balancing the risk against the gains.

Unluckily for Mr Camm, one of the overseas biologists working on the Reef, Professor Joe Connell of the University of California, had given a public lecture the night before on the Santa Barbara spill. The oil-slicks at Santa Barbara, said Professor Connell, had spread over

100 miles, and six-and-a-half months after the first leakage oil was still pouring out and the oil companies' best efforts could not stop it.

Did Queensland want an oil industry there, or the Reef as it was? It could not have both.

He felt the real danger was less from oil spills than from the mere entry of the oil industry into a tourist and scientific area such as the Reef. It would mean dredging for pipelines, drilling, mud dumping, other kinds of pollution as well as oil – and the case of Louisiana showed that continual oil spill and other problems would most likely follow.

This was reported faithfully beside the Mines Minister's statements. As Connell had worked on the Reef itself for several years, as well as knowing the Santa Barbara situation at first hand, his credentials could hardly be questioned.

The Great Barrier Reef Committee, however, had not yet officially revised its views on 'controlled exploitation'. Instead, Dr Patricia Mather produced a new proposal for a Commission or Authority to be set up by the Commonwealth to 'watch over the Reef', with subcommittees to advise it and recommend legislation for it to prepare. It should control all Reef 'exploitation' – including, we sadly noticed, mining – and 'zone the Reef' according to the different purposes of the exploitation and their acceptability.

Now came the state Liberal Party Conference. On the first day John Gorton was reported as having said he was 'strongly opposed' to oil-drilling or limestone mining on the Reef. It was his 'personal desire' that the Reef be left the way it was. (This was, it turned out, a statement made to the press and not in the opening speech.)

The State Ministers at the conference were angry and agitated at the report – naturally a front-page headline. They said that the whole matter would be discussed at the Monday state cabinet meeting, and a report might be released on the Commonwealth's own involvement in negotiations for offshore leases.

The Prime Minister had said that the Commonwealth's powers over the Reef were 'fairly limited', especially in the case of leases granted before the question of Commonwealth control beyond the three-mile limit had come into consideration. But he said he would 'do whatever he could' to see that the Reef was protected from pollution or ecological disturbance. (This statement, widely reported all over Australia, lulled some people into believing that the whole question was going to be solved without more public agitation. No doubt with the best of intentions, the Prime Minister had thrown cold water on the very public concern that was working for the Reef's survival.)

Mr Gorton added that the Commonwealth had been ready to discuss the Crown of Thorns problem with the state government since July. It was up to the state to arrange the meeting. He left for Canberra, and behind him he left dismay and disarray.

Perhaps Mr Gorton had hoped to sway the Liberal Parliamentary Party towards overruling the Country Party on the Reef. If so, he reckoned without the strength of the long-entrenched Mines Department and the attitudes of the Minister, the Premier, and the other Country Party members of the cabinet. We were later to get some insight into the reasons behind their confidence in continuing with the drilling programme; but confident they were. They had no fears for the Country Party in Queensland in the coming federal election, as no doubt the Liberals had for their own federal votes. They had taken their stand and they would stick to it.

The Mines Minister declared that every prospecting permit issued by the state government had been endorsed by the Commonwealth. (Indeed, under the terms of the Offshore Agreement, we knew, this had to happen.) The Commonwealth was already allowing the Tenneco drilling in the Gulf of Papua at the northern end of the Reef area, and several drills had gone down, with gas shows in some. These drillings were 'the sole responsibility of the Commonwealth'. If the Prime Minister was prepared to take responsibility for any claims for damages, he said, he was 'sure the Queensland government would be prepared to discuss his suggestions on how to cancel the permits'.

At the remaining day of the Liberal conference, the state Liberal leader, Gordon Chalk, was reported to have managed to 'take the heat off'. He appealed to the convention to discuss the motions on the Reef but not to vote on them; a vote 'might damage the Liberal Party'.

The convention debated for forty minutes, but the only resolution passed was an off-the-cuff one, simply calling for 'extreme care' to be taken on the Reef.

On Monday the Mines Minister repeated his Saturday statement, and asked complainingly why there had been 'such a fuss' about the Repulse Bay drilling, when several drills had already gone down on the Reef. (These were, of course, before the Santa Barbara disaster.) This got a furious response – and we, incidentally, scored another country branch from it.

That week, answering a question from the Labor side in Canberra, Mr Gorton once again repeated his view that the Reef should not be put in danger by oil-drilling. The Labor Party asked the Commonwealth to introduce legislation to remove dangers to the Reef. It also asked what the position was about Queensland's right to grant the Reef drilling permits. If Queensland's jurisdiction might be found to end at low-water, had the Commonwealth the power to rescind the permits?

The Attorney-General, Mr Bowen, answered that the leases granted could be partly within waters under Commonwealth control; but under the terms of the Offshore Agreement, they were 'to be continued'. They were, he said, only 'permits to explore'. The bulk of them were well beyond the three-mile limit claimed by the states, but in some cases they might extend within it, particularly off islands.

The *Courier-Mail* commented that the Prime Minister's instincts were right if he believed that the Reef issue was of great significance to the electorate at large. 'Conservation and ecology have become the "in" words of the day. And so they should be.'

'The Labor Party, both state and federal, is going its hardest to make political capital as the gallant defender of the Reef,' said the *Courier-Mail* disagreeably. This was true; but we were sure that Whitlam was in fact really interested in the Reef's fate.

He had asked in Parliament what consideration had been given to the ACF suggestion for a special inquiry and the setting up of a joint commission. The Prime Minister had said that this was being considered; but the matter was not just one of setting up a commission. It involved the question of what authority the commission was to have. And he said he might order a complete Barrier Reef study.

John wrote that he was 'absolutely delighted' with this. 'I put a lot of work into that one. I would say that he could now be safely said to be very clued up on the real facts of the Reef, and knows what he is talking about. Les Arnell and I will be writing to him, raising further legal objections as to the validity of the Queensland government's leases for drilling. I believe we have quite a few interesting points to make to demonstrate that the Commonwealth *can* stop drilling on the Reef.'

A case had just been decided by the High Court of Australia, which had far-reaching implications for the question of Commonwealth versus states rights in the territorial seas. It had begun as a minor prosecution of a fisherman, Bob Antonio La Macchia, who had been caught by a fisheries inspector six-and-a-half miles east of Barrenjoey in NSW, catching fish with a mesh that was smaller than prescribed under the Fisheries Act. He was prosecuted at the Sydney Court of Petty Sessions; but the question of legal jurisdiction in the case had arisen, and the Attorney-General had removed the case from the Magistrates' Court to the High Court because of the important constitutional aspects the case raised.

The High Court judgement was to be of crucial importance in the question of offshore jurisdiction. Its ruling was that the states had no rights or jurisdiction over territorial waters adjacent to their coastline, nor any rights or jurisdiction over the seabed. This meant also that the states had no separate rights to the minerals or other natural resources of the seabed. The only rights to require royalties or to legislate in these areas lay with the Commonwealth. Any state legislation dealing with offshore oil or any other natural resources therefore had to be four-square or 'mirror' legislation with that of the Commonwealth.

This decision, with its implications, must have caused a fine panic among states Attorneys–General. But its effect on states legislation had not yet come into the open, though probably the decision – in which Sir Garfield Barwick was involved – influenced Mr Gorton in the stand he took towards the Reef.

John's Innisfail firm of solicitors now drafted a letter to the Prime Minister, which John signed as president of our Innisfail branch. The letter suggested that as a result of the High Court decision, it would appear that any licences either to explore or to drill for oil in offshore waters, granted by the Queensland government before the intervention of the Commonwealth, were in fact invalid. It was also doubtful whether the Commonwealth had the constitutional jurisdiction either to delegate or share its supreme authority with the state of Queensland, since the state had no existence in this matter in the eyes of international law.

The Commonwealth could not, for example, grant Queensland the right to go independently to war with Indonesia. It would also appear that there must be considerable doubt as to the ultimate validity of that rather hasty piece of legislation, The Petroleum (Submerged Lands) Act of 1967.

As with the recent case of R. v. La Macchia, which, because of its important constitutional aspects, was removed ... from the jurisdiction of the Magistrates' Court, we would suggest that a similar course of action be followed in the matter of drilling licences granted by the Queensland government to Ampol-Japex in Repulse Bay.

If such a course of action does not appear possible, then there would appear to be only one last, desperate resort to prevent the eventual and inevitable destruction of the Barrier Reef by oil pollution, that is, the issue of a writ against the Queensland government and Ampol-Japex by a combination of such societies as the Wildlife Preservation Society of Queensland, the Queensland Littoral Society and any private individuals who care to join in issuing the writ. This, of course, would throw an intolerable burden on these individuals and small societies with limited funds, and it would not redound to Australia's credit in international circles that such methods had to be employed to save one of the seven wonders of the world from completely unnecessary despoliation. However, a sustained Australia-wide public campaign for donations might make available sufficient funds for legal expenses to carry the case to the High Court and, if necessary, to the Privy Council.

In the meantime, a total voluntary public boycott of all Ampol-Japex products is being considered.

This idea had come from our Gold Coast branch, which early in August had paid for an advertisement in *The Australian*, calling for such a boycott. This had been taken up in a number of letters to the press, and letters to Ampol from various people in these terms had, we knew, followed it.

While we waited for some reply to John's letter, we turned over the idea of asking all our branches to join in a series of advertisements of the same kind. But we decided to postpone this, and instead to write to Ampol's chairman. The letter read:

You are undoubtedly aware of the great and mounting concern for the future of the Great Barrier Reef and all its associated reefs and coastlines when oil-drilling procedures commence shortly under your Company's farm-out agreement with Japex in the Repulse Bay area.

We understand that under the farm-out agreement conditions, drilling operations must be carried out by Japex.

We know that your Company is proud of its Australian public image. At present, there is a very strong feeling that this public image is being damaged by the proposed activities in these highly vulnerable and important tourist areas ... We are therefore taking the step of appealing to you and to your Company to withdraw the conditions of your farm-out agreement and to state publicly that this is being done in the long-term interests of Australia as a whole.

A favourable answer would be the greatest advertisement that at present your Company could have.

We decided to give Ampol a chance to reply to this, before going ahead with the wider appeal for a boycott. In the event, no reply was ever received.

But more and more people now felt that action was up to the Commonwealth Government. In an article in *The Australian*, Ian Moffitt said that 'only bold intervention at the highest level' would save the Reef from the shadow of oil towers. Conservation might be becoming respectable, he said, but it was having little or no effect on government action.

The Great Barrier Reef Committee took the opportunity of the general public disquiet with government attitudes to the Reef, to reveal that the biological research facilities at Heron Island were being starved for equipment and support. The state government grant of $2,000 a year for maintenance was not enough to cover its running costs, and scientists were 'being forced to beg for funds'. Would the Commonwealth and the state accept the responsibility? Heron Island was 'one of the best sites for a marine research centre in the world' if it had the right facilities.

August had certainly been a big month for Reef publicity; but not, it seemed, for action. There was no sign of a halt to the Repulse Bay drilling. The chairman of Exoil wrote to the Press lauding the merits of Corexit, the new 'non-toxic' dispersant in which the oil companies were now putting their faith.

However, we knew from Professor Connell's lecture that its use had been prohibited at Santa Barbara, and he had said that the claim that it was non-toxic to marine life was incorrect. We wrote to contradict Mr Siller.

We had felt that we needed a solicitor in Brisbane who would be willing to work on the Reef question, as the Innisfail firm was doing, and in that same month we found one. A solicitor stated to the press that a paper on the Continental Shelf question, given at a legal convention in Brisbane in July, had seriously questioned the state government's power to grant prospecting authority to Ampol over the area it had applied for. This area stretched for a hundred miles each side of Mackay and seaward for forty miles. He considered, on the basis of a case of his own, that it might be held that any distance over three miles from the mainland was in 'international waters'.

The Mines Minister replied that the exploratory drill would be in Queensland waters. (In fact, the first drill was to be within the waters of Repulse Bay which did, in one definition, come under Queensland jurisdiction – that is, within a line stretching from headland to headland of an enclosed bay.) But he did not go into the question of the state's authority to grant the permit for the whole area – which stretched far beyond even the three-mile limit.

I went to see the solicitor and showed him the letter from John to the Prime Minister.

He was cautious. He was not well up in international law, and would have to have a look at the situation. As for the idea of a writ or injunction, he said that 'if it looked possible, he would have done it himself'.

The question was, he thought, whether the States Mines Ministers, in their agitation over the question of states jurisdiction, might have forestalled any action in some way. If they had, this approach would come to nothing. Not only Queensland was affected by it; the question of the legality of leases issued by every other state government in offshore areas would be raised. The royalties going to state governments such as Victoria, from the Bass Strait leases, would be in question. Could the Commonwealth afford to risk making this claim, with the consequent unpopularity they would gain in the states – and just before an election?

We had thought of this too. Already Gorton was being attacked by the states for his 'centralism'. It would be a dangerous ploy for him to raise the constitutional issue, particularly now. We were sure that pressures were on him, not only from Queensland, to delay the proposed offshore mineral resources legislation, of which nothing more had been heard.

His government was not as popular as it had been. The Vietnam War was embittering many people; there was division within the government itself, and jockeying for position. The Great Barrier Reef meant votes but it also meant opposition from the states whose royalties and prospective royalties from offshore mining and drilling were threatened. Even more important, it meant opposition from the mining and oil industries – and we knew how strong those interests, many of them international, were when their prospects were in question.

We could not, in fact, be sure which way the Commonwealth Government would jump, either on challenging the states over their legal justification. for issuing permits, or on the question of Barrier Reef protection. The idea of a moratorium, which we knew had been seriously considered, seemed to have been dropped. That was indicative of the opposition we were up against.

Moreover, it might well be that unless the Commonwealth did step in and claim all rights, a successful fight against the Queensland permits might merely result in Reef waters being claimed as 'international'.

Since many of the companies which held drilling permits were overseas-owned, we might be worse off then than before.

The result of the federal election, then, would be crucial. If we could prove that enough votes were indeed to be won, on the Great Barrier Reef question, the government, whichever party it might be, would be much more inclined in the direction of cancelling the drilling.

Meanwhile, the International Union for the Conservation of Nature was much concerned over the way things were going. Its World Wildlife Fund – which we had approached both via Alison Büsst and on my visit to the headquarters at Morges – made an unofficial approach suggesting finance for an international biological research team to visit the Reef. This was to come to nothing, since the Great Barrier Reef Committee had already made a report to the Queensland government on needed research, and the Academy of Science was known to be interested; but it demonstrated the amount of international concern, and we were grateful for it.

Our Gold Coast branch had the next inspiration. The local Show was on, and they were running an exhibit on the Reef. Now they decided to run an impromptu public opinion poll among the visitors to the Show. They provided a ballot box, with voting forms – for or against drilling on the Reef. The result was cataclysmic. The number of adults voting against drilling was 976, for drilling 2, with a few informal votes.

We had known that there was strong opposition; but this was amazing. An opening was obvious for a Queensland-wide sampling of opinion, if we could organise this – and our branches were one avenue for doing it. John wrote in high excitement; the Innisfail Sugar Festival was about to be held, and already the local branches of the society and the Littoral Society were organising a display on the Barrier Reef. They would hold a similar poll at the same time.

We set about devising a questionnaire that we could use all over Queensland. It had to be one that would be watertight in its wording, so that no hidden bias would weight the results, and we submitted it to psychologists to make sure that this was so.

The Liberal Party conference had left sores in many Liberal supporters really concerned for the Reef, and no doubt some too in those who were merely worried about the party's election image. There were letters to the papers, when it became clear that the government was going ahead with the Repulse Bay project, condemning the handling of the conference and the smoothing over of the real issue in Mr Chalk's account of the conference's outcome. The indignation felt was to have a very useful result. A wholly new organisation now entered the field – one not connected with the conservation groups except in sympathy over the Reef.

Until then, the main fighting and publicity jobs had been left to the conservationists, though there were so many letters to the paper from people otherwise outside the conservation battle that it was evident that people far beyond conservation groups were in agreement with us. But there was a steady campaign to rubbish conservation groups and their members – we had been described as, in turn, agents of the American oil companies and anti-Japanese (since we were opposing the Japex drilling), and as communists determined to keep America from getting needed minerals for the space programme from beach-sands, since we opposed sand mining in the Cooloola area. It was therefore a big relief in many ways when after the Gold Coast poll and the Liberal conference this quite independent group, mainly of Liberal and even Country Party persuasions, started a citizens' Save The Reef Committee. We had had no contact with their members up to now, and the formation of this new group took some of the onus off the conservation groups and demonstrated the real concern of the public.

The Save the Reef Committee had its first meeting in mid-September. I was asked to attend. The meeting was chaired by Senator Georges, a Queensland Labor member of the Australian Senate; while one of the members of the committee was a state Liberal backbencher. I was asked to be patron of the committee, and the members began discussing what they could do to help the campaign for the Reef.

It would be a godsend to us to have more hands, and another organisation, on the job, with the proposed Queensland-wide poll coming up. We were already arranging for the poll forms to be printed, and corresponding with our branches – of which we now had twelve – to tell them what to expect and to ask for their co-operation. We were terribly overstretched, both for time and money; our then secretary lived many miles from Brisbane and some of the secretarial work fell on me. We were not looking forward to the prospect of having both to conduct the Brisbane poll, with the Littoral Society, and organise the results from the rest of Queensland.

The new committee, after some initial doubts as to whether they would really get a vote of enough strength to influence political thinking, were convinced by the Gold Coast vote and agreed to help with the balloting in Brisbane. Now we had more people to conduct this, and our hardworking Brisbane members could relax a little. The society was in any case organising the printing and distribution of the forms – a big enough job in itself.

I had, for once, some acceptable qualification in conducting opinion polls. My university courses had included psychology and anthropology; and I had once been employed by an advertising agency in preparing poll surveys and questionnaires and analysing the results. Also, I had once held the position of statistician at the University of Queensland. We were sure that the questions were unweighted; the problem was that the interviewers we could use were quite untrained in the job, and their own views might be known by those who were questioned.

We had to concentrate on getting so many people polled that the result would be beyond doubt; I reckoned that we would need perhaps 4,000. And the polls would have to be random – in shopping centres, by door-to-door interviews, or wherever we could find a lot of people whose sympathies might be expected to be representative. I had drawn up an instruction form that emphasised that branch members and known sympathisers were not to be polled.

Since we wanted the results in, counted and checked, before the date of the Federal election, we asked that the poll should be done as quickly as possible. The forms were to be either signed or initialled by the person polled, so as to avoid any possible accusation that the poll was 'cooked' or unproven. Each area was to have at least 100 people polled, the more the better.

Since we hardly expected to get all the polls in before the October election date, we decided to publicise any important trends as the main results came in.

Meanwhile, John came to Brisbane, armed with the Arnell opinions and the copy of his letter to the Prime Minister, to consult our solicitor about the question of the legality of the Queensland drilling licences and the possibility of issuing a writ. The solicitor had agreed that if the Mines Department was not specifically authorised by the Commonwealth to act on its behalf, we could indeed move. The question was whether this had been done.

There was to be a meeting of all States Mines Ministers, to consider the legal position. There was much at stake for them all in the question of offshore exploitation.

Our solicitor set to work to discover whether the Mines Ministers had in fact now become legal agents of the Commonwealth by being appointed under Commonwealth legislation.

On 26 September he advised us that 'in fact, just this thing has been done. In view of this, quite frankly we cannot see that any legal move can be made to get the present arrangements declared invalid'. That seemed likely to be the end of that. But we kept the idea of a possible writ, to be issued on other grounds, in our minds. The problem would be money – lots of money – and we were too involved in the public opinion polls to run an Australia-wide appeal for the funds needed, with all the extra work that would mean. Later on, we might have to do this; at present, a clear demonstration of public opinion against the oil industry getting established on the Reef was the main object. It would be not only an electoral advantage, perhaps for years to come, but a weapon that might have other important results.

Meanwhile, the Innisfail conservationists went ahead with their own poll. They had had to have the forms printed before we decided on our own wording, so that it could not be counted in our results (though they faithfully conducted a second poll for us when our own forms arrived). Also, they ran a float in the festival procession – a black coffin surmounted by a model oil rig – and it is conceivable that this might have influenced the poll. The results were just as striking as the Gold Coast one had been – 517 people were against drilling and only 6 for it. That, too, got plenty of publicity.

The Save the Reef Committee was circularising all Federal candidates, asking them to express publicly their personal attitude to oil-drilling (not merely the party line). The unofficial polls were certainly helping politicians to see which way the wind was blowing.

On a suggestion made by a friend, we now wrote to the owner of the *Yomiuri* in Japan who was also a major newspaper proprietor. Our letter pointed out that there was no doubt, after the Santa Barbara tragedy, that oil would injure the Reef both environmentally and aesthetically.

The Premier of Queensland has stated oil-drilling must go on, because the permits have already been granted and cannot be rescinded without great expense to Queensland. We know that the proposed drilling in Repulse Bay by Japex Ltd under a farm-out agreement with Ampol Petroleum Ltd was arranged before this concern became obvious, and that the original mistake did not lie with your oil-industry but with our own government ... It is our view, and that of many others, that all proposals for oil-drilling on and near the Great Barrier Reef should be abandoned in the long-term interests of the world itself, as well as of Australia. We will be most thankful for any help that, at this stage, it is possible for you to give to our objectives.

We were getting so much public support, including that of women's organisations in Queensland, that things looked good. But the Australian Conservation Foundation was standing on its laurels so far, making neither any further comment nor, as far as we knew, any follow-up approaches to either the state or the government. Their annual general meeting on 15 October was approaching. Elections for the Council, for which both Des Connell of the Littoral Society and John Büsst had been nominated, had resulted in Des getting in, but not John.

We wanted an approach to Ampol from the ACF on the lines of our own letter; and I had another bit of publicity I wanted distributed to the press while I was in Canberra. I had written to one of the most famous overseas zoologists and conservationists, Professor Grzimek of Serengeti fame, who ran a television programme of great repute in Europe. I had told him of the oil-drilling proposals, and asked that he write to the Prime Minister and to the Premier of Queensland with his views. He responded with an excellent letter.

He had sent me a copy of this, and though it was addressed to the governments concerned, had authorised me to use it as I wished. The Foundation meeting seemed a good time to release it to the press. I hoped that it might also galvanise the ACF into some further action.

I have read in German newspapers of the decision of the Queensland State Government to allow drilling for oil in offshore areas of Queensland where the whole future of the Great Barrier Reef and of Queensland's still largely unspoiled northern waters and beaches is at stake. The recent case of the Santa Barbara oil blowout more than sufficiently demonstrates what will in all probability be the fate of this great wonder of the world should oil be found; and even if it is not found, the pollution consequent on any drilling activity will in itself constitute a threat to the marine ecology of the Great Barrier Reef.

Besides my work as a Director of a Zoological Garden and at the University I do a permanent television programme in Middle Europe with a permanent audience of about 40 million people. In this programme I had already reported on the wonders of the Great Barrier Reef. To destroy or endanger it would be a similar blow to the civilisation of mankind like the destruction of the Acropolis or St Peter's Cathedral in Rome.

The ACF did not have anything relating to the Reef on its meeting agenda and the open session was very brief, because of a book-launching to follow. After the Chairman's report, there was no time to discuss the Reef situation.

But I had time to show Sir Garfield privately the letter we had written to Ampol. He did not think the ACF would agree to writing a similar letter, but would await the response to ours and to a private approach. The states-Commonwealth relations were bad, and until the election was over he thought the ACF should refrain from further approaches.

But we managed to get John Büsst invited onto the Council as one of a number of additional members nominated. We now had a strong contingent on the ACF Council.

And the Grzimek letter was given high publicity in the press, when I had handed it to a young reporter friend who took it over enthusiastically. We had achieved something, at least.

The next meeting of the Save the Reef Committee adopted the poll circular and forms we had produced, and the work of polling in Brisbane began at once.

By 22 October we had polled 2,400 in Brisbane, at shopping centres, at railway-station entrances, and by door-to-door interviews. The result was as good as we could have dreamed of. Ninety-two per cent were against drilling on the Reef. We got this information out as soon as we could to the press and to politicians. Since the Prime Minister had already declared himself in favour of stopping the drilling, and the Labor Party policy was similar, this did not mean that either party got an obvious advantage from the polls; but it did, so we thought, confirm that Mr Gorton's views – which were, we knew, not universally popular in the Liberal and Country parties – were approved by the electorate.

The government scraped back in, narrowly. 'Maybe now the politicians will vie with each other to save the Reef,' John wrote.

A new branch of the Littoral Society had been formed in Cairns; and the Cairns poll had been carried out. In an area which might benefit from some employment by the oil industry in their Reef projects, we had 239 against drilling and only 33 in favour. It looked as though we were going to find much the same trends everywhere. 'With evenly divided (Commonwealth) government forces, these public opinion polls could have considerable force,' John wrote.

It was some time before we got the final results of the poll, since some country centres were slower in getting in the results. Our office was stacked high with completed poll forms, all of which had to be recounted and checked for validity. But we were releasing progressive results as they came in, and the press printed them.

The final results showed that of slightly over 5,000 people polled, a total of 4,646 were against drilling and only 285 in favour. We had been triumphantly justified in thinking that the feeling against the establishment of an oil industry on the Reef was overwhelming. All the country centres had shown the same trend; and this was important for politicians considering the feelings of their electors. We hoped the polls would, indeed, be influential.

But in the meantime, the Ampol-Japex operation was still being prepared. Drilling had been due to start early in October; but the drilling vessel had been delayed in setting out. It was being fitted in the United States, with what the state government said was the latest and most foolproof equipment, which would, according to the state Mining Engineer, 'almost eliminate' any danger of a blowout or oil spillage. New regulations had been issued on safety precautions, and a Mines Department official would also be stationed on the rig on a twenty-four-hour basis to superintend the operation, as the Mines Minister had promised.

Meanwhile, other equipment was being assembled at Mackay and there was no sign that the operation would be stopped. The company had not set a date for the drilling, because of surprising delays in converting the rig – it was being leased by Japex and was a former US Navy vessel.

Now an economist from the University of Queensland came into the argument. Speaking to a public meeting organised by the Save the Reef Committee, he said that the supposed advantages, listed by the Mines Minister, of discovering oil on the Reef, were highly questionable. He pointed out that in fact a big oil industry in Queensland would raise the price to the consumer – as had already happened with the Bass Strait finds; and the oil royalties charged by the Queensland government at present meant that the people of Queensland were providing the money that the state government now got from its Moonie field. Any oil discovery meant that petrol prices to the consumer were raised; and of a two cent rise, perhaps one cent would go to the state government in royalties. The rest would go overseas.

The tourist industry, he said, was already a big money source and would become more so. As for the question of compensation to companies already granted drilling permits – which had been quoted by the state government at a minimum of ten million dollars – even if this were a correct figure it would not be much to pay to safeguard the tourist industry. He estimated that Australia was already seventy per cent self-sufficient in oil.

Dr Mather, the Great Barrier Reef Committee secretary, had written to the press in September with a new proposal for a Great Barrier Reef Commission – not one, as the ACF had advocated, with advisory powers only, but with power to make and enforce decisions on the Reef's use and with strong biological advice. This got a lot of support in letters; and we ourselves agreed that it was a necessary and promising proposal. The Committee had been disillusioned by the state government's ignoring the Endean report on the Crown of Thorns starfish plague, the Ladd Report, and the lack of research money, and apparently now realised the need for some Commonwealth involvement if anything was to be done on the Reef's problems.

In fact, the Federal government had just launched the first prosecution under its new Act giving it power over the Reef's living resources. This was against a Formosan clam-fishing boat. The government was demonstrating its goodwill in enforcing the new legislation.

Meanwhile, the Japanese petroleum exploration company was growing embarrassed by the extent of the protest against the Repulse Bay project. There were rumours that Japex might be willing to withdraw if it was suitably compensated. But Ampol, and the Premier, denied that there would be any moves to stop drilling. The vessel was still delayed, and we began to hope that, if the delays continued long enough and the controversy loud enough, we might yet prevail in some way or other.

In Tokyo, the manager of Japex's overseas department admitted that the programme had run into 'delicate problems'. The Prime Minister's open opposition to drilling on the Reef was no doubt one of these; but also the Japex employees in Australia had been surprised by the hostility they were encountering.

Mr Gorton announced, in his pre-election speech, that an Institute of Marine Science would be established at Townsville. This caught the imagination of many, beyond marine scientists. The amount of money so far spent on marine biology by Australia was almost a farce; so little was known about the Reef by biologists, in comparison with geologists, that the proposal was an exciting new step.

Mr Gorton tied the proposal in with the world's new interest in the oceans and what they could produce. In August there had been an important United-Nations-sponsored conference on the oceans, attended by a Commonwealth mission. An inter-departmental committee had been set up by the Commonwealth to discuss ocean science and the policy for ocean research in the years ahead. An Institute at Townsville, if it was indeed to be of 'world excellence', as the Prime Minister promised, would be a very important contribution to world scientific knowledge, and its base – the Great Barrier Reef – would be a marine laboratory that could not be equalled anywhere.

John Büsst and Professor Burdon Jones were delighted with the promise; but there were signs that the state government might not be. We wondered about the attitude of the Great Barrier Reef Committee scientists, who had continually pressed for an upgrading of the Heron Island station.

And the Mines Minister would clearly have no part of Commonwealth interference on the Reef. Addressing a Save the Reef Committee lunch meeting, he said, *inter alia*, that 'Australia was in danger of losing the Reef to overseas interests'. The question had been raised overseas of the rights of countries to exploit areas of the sea beyond their territorial limits. If Australia's Barrier Reef were to come under the control of the United Nations, he said, will fight tooth and nail to retain the Reef, and so will the Commonwealth Government. We in Queensland claim the Great Barrier Reef, and believe we own it. If Australia had to fight for its right to the Reef, it would help if it could show now that it was doing something to develop it'. And this apparently meant the drilling and exploitation of oil on and near it.

This strange claim set off yet another round of protests. *The Australian*, in an editorial, wrote:

Conservation this week took a historic step forward … Mr Camm revealed that the only way to save the Great Barrier Reef for Australia is to drill it full of oil-wells with the greatest possible speed … Mr Camm's prescience in approving oil-exploration permits means, fortunately, that no time need be lost in putting the defensive plan into action …

The next action must be increased naval patrols to prevent international poaching of our prized Crown of Thorns starfish, nurtured so carefully on the reef by skilled governmental vacillation.

We joined in with a letter:

Queenslanders, and for that matter Australians, have long wondered when our Mines Minister would either answer any of the arguments against establishing an oil-industry in Great Barrier Reef waters, or provide a convincing argument for doing it.

His recent remarks, while they don't attempt to reply to the arguments *against*, certainly provide a fascinating display of his powers to argue *for* the Reef's destruction ... We may not, in international law, *own* the Great Barrier Reef, but at least we are far and away the likeliest people to make a really fine job of wrecking it.

And now, to our delight, the work John Büsst had been doing with the trades unions began to show its first practical results. He had helped draw up a submission for an Innisfail member of the Amalgamated Engineering Union to present to the ACTU Conference. This was to be published in the union journal; and the AEU had endorsed the campaign to prohibit mining or oil-drilling on the Great Barrier Reef. Mr Bob Hawke had given the submission his own attention; copies of it were to be sent to the Prime Minister and the Premier.

The concluding passages of the submission as presented were:

It is therefore resolved:
1   That a total ban on all mining on the reef be immediately declared.
2   That an independent scientific and judicial commission be set up to determine the future of the Great Barrier Reef with power to co-opt all such international scientific assistance as thought necessary.
3   That the Commonwealth Government be requested to issue an originating summons to the High Court to determine the constitutional issue involved.
4   That a writ be issued against the Queensland Government to prevent its proceeding with the decision to allow Ampol-Japex to drill in Repulse Bay.
5   That a public opinion poll on mining on the reef be conducted at every major centre on the Queensland coast before the Federal elections. [This we had ourselves pre-empted, of course.]
6   That the Barrier Reef be declared a National Marine reserve for the benefit and relaxation of the Australian public, to be in no way despoiled by the activities of mining companies.
7   That a voluntary Australia-wide boycott be called for on any oil or mining company endangering the future of the Great Barrier Reef by mining operations.

This was to be a huge, indeed crucial, step forward. A black ban on all mining and drilling was the only thing that would really and decisively prevent it; it was also something that the Queensland government had no power to prevent. It was not yet official ACTU policy, but the Amalgamated Engineering Union had endorsed it, and they were among the most powerful unions associated with mining.

Working as we were not only on the job of collating and publicising the public opinion polls as the forms came into the society office, but on a number of other problems as well, we in the society were enormously cheered by the news.

Now, it remained to be seen what would happen.

7

The Unions Move In

Meanwhile, the Crown of Thorns controversy raged on. Early in September, a diver and underwater photographer named Ben Cropp told the press and the Premier that the extent of the plague had been seriously exaggerated. Other divers, and a research biologist, supported this view in varying degrees. The 1968 Endean report, with its request for a diving team to control the starfish by hand collection and to try to confine it to the area of the original outbreak, and its recommendation for a ban on triton shell collection, had not been implemented by the state government.

The Premier now lent a willing ear to Mr Cropp's assurances. He announced that he was rejecting the Endean report. 'Dr Endean's qualifications do not establish him as an authority in this matter.'

The facts that Dr Endean had been commissioned to make the report and that he was the only Australian scientist working on the problem did not appear to weigh with him. Nor did the fact that Mr Cropp was not a marine biologist.

The Premier's views were convenient, in that the state would not have to spend money on a diving programme or on further research. They were also convenient to the Reef tourist industry, which was still worrying about the publicity given to the plague, which might interfere with the flow of tourist money.

Endean now took a six-man team to the Reef near Innisfail, to survey the situation and see if predictions of the spread of the plague had been fulfilled. He came back with alarming reports. He had found only five reefs between Irmisfail and Townsville not affected, and huge coral formations 'hundreds of years old' had been demolished. He was sure that the plague was not a cyclic occurrence; the age of these formations and the extent of the plague pointed in the other direction. But the state government seemed unmoved.

And few people, even now, realised the full extent of the oil-drilling applications. Finally, after many requests, the Minister for Mines released the map of offshore areas held under title as at 4 September 1969. Those who studied them were taken aback. The only substantial area in which no permits had been issued – and presumably the only one in which no company expected to make a strike – was near Cairns.

The main emphasis of much of the press publicity went to the question of whether or not oil would harm the coral. It was the living coral of the Reef, so spectacularly beautiful in its shapes and colours, so photogenically displayed in tourist posters and advertisements, that people thought of when they thought of the Reef – that, and the fact that the Reef itself had been built by the coral polyps. If coral could survive oil, they reasoned, then the Reef itself would survive. The staggering complexity of the Reef community with its 'hundreds of thousands of species', of which Dr Talbot had spoken at the ACF Symposium, and the lack of knowledge of how these species interacted, bred, fed, and replenished the Reef, was more difficult to grasp. The Reef was a coral reef, to most people, and that was that.

Yet no one knew, or could make a guess, at what the effects of oil on this total community would be. If the other species were affected, could the coral itself survive? What were its own requirements for life? What organisms, and what conditions, did it depend on? And there were many species of coral, possibly each with its own special food requirements, its special relationships and adaptations. Almost all that could be said about the Reef was that nothing could certainly be said.

The Crown of Thorns outbreak at least indicated that something had gone wrong, somewhere, with the almost unknown food and predation chains of the Reef. Some predator, or predators, perhaps was missing, or some factor had encouraged the Crown of Thorns to breed and feed wildly and spread in thousands where before few of the starfish were found. And many biologists were coming round to the view that human interference of some kind, whether through pollution or depletion of other species, was responsible.

But plenty of people were unable to believe that the Reef, so huge in its size, could not support all kinds of interference. Surely, in a structure so vast, some mining, some drilling, some shell collection, some pollution, could make little difference?

The University of Queensland geologist who had already supported Rhodes Fairbridge's views was one of these. 'The Reef,' he said in a letter to the press, 'is big enough for all. Let part be reserved for nature, part for commercial exploitation.'

This was the point that ecologists like Dr Grassle disputed. Giving evidence to the Senate Select Committee on Offshore Petroleum Resources, he argued that the very size of the Reef could well be the factor that allowed it to have so many species.

Dr Des Connell, of the Littoral Society, in an address to the Queensland Petroleum Exploration Association in September 1969, said:

> In extensive studies of pollution all over the world, biologists have been seeking biological indicators of pollution. It now seems that the best indicator of changes in biological systems is the index of diversity. The diversity of species on the Reef is much higher than in any other aquatic environment. For example, the reefs of the Caribbean have only about 40 species of coral, whereas the Barrier Reef has at least 200. Dr J. F. Grassle found over 100 species of polychaete worms in a single 13-pound lump of coral – this is more than the number collected in extensive surveys over long periods of time in other regions. Every instance of marine pollution has resulted in a reduction in the diversity of plants and animals. The Great Barrier Reef is more susceptible to pollution than other environments because even the slightest changes in the environment can result in tremendous reduction of diversity. The reason for this is that all the species of the Reef are highly specialised in their requirements … Any slight disturbance of such a system could have tremendous consequences.

It began to seem that the Crown of Thorns outbreak could be the first noticeable 'tremendous consequence'. But the state government did not want to hear about it. It sent a team of three police skin-divers to study the starfish at deep levels; and took no other action.

With both these arguments raging on, there was seldom a day without some news item or article on the Reef and its problems. Piling up my newscuttings in a large box, I thought bitterly that there was only one danger to the Reef, and that nobody had yet pointed it out. The danger to the Reef was civilisation; and if the Reef was sick, that sickness was caught from us. But no one was suggesting research into people.

Meanwhile, the reports that Japex was hesitant about going on with the Repulse Bay drilling seemed to be coming to nothing. The letter we had written to Ampol had so far been unanswered. If anything had come of the private approaches we had heard about, there was no indication of it. Only the continued delays to the arrival of the drilling vessel, the *Navigator*, from Texas, might, we felt, give a clue to moves behind the scenes.

The threat from limestone mining, at least, seemed to have receded. We had been interested in May to notice that the Minister for National Development, Mr Fairbairn, had told the Commonwealth Parliament that the Queensland government had made it clear that it would not allow mining on the Reef. The Ellison Reef case had not been in vain.

The Gorton government was still in strife. The election results had been so close that the Prime Minister's leadership was being questioned even by his own party. Some of his policies were under strong attack; the Labor Opposition kept hammering weak points, and the unrest over the Vietnam War was causing the deepest division the country had ever known. While this unrest was largely confined to the young, at least at the beginning, many older people, even the staunchest of Liberals, were now involved.

The near-final results of our public opinion poll in Queensland were published in the press, with good publicity, at the same time as one of these challenges to the Prime Minister was being made. It was not only we who felt that one of the Prime Minister's policies that did meet with popular approval, and was not being openly challenged anywhere except by the state government and the oil companies, was his stated opposition to oil-drilling on the Reef.

But Queensland's Premier still took the opposite stand. He was asked in the Parliament what attitude he was taking to the expressed 'strong objection' by the Prime Minister to drilling for oil in Reef waters.

Would the Premier ask Mr Gorton if the Federal Government was prepared to underwrite any possible compensation claims if the drilling were halted? And if it agreed to do this, would any further offshore drilling be banned by the state government?

The Premier answered that the state and Commonwealth governments were equally involved. They had jointly entered into statutory agreements, and the state was not going to repudiate those undertakings. 'If the Commonwealth government decides to exercise powers to prohibit drilling offshore, that is a matter between the Commonwealth and the companies and persons affected. My government will not be a party to repudiation.' He went on to reiterate that the strictest conditions would be imposed on any offshore drillings, which would make the risk of a blowout remote and the risk of major damage to the Great Barrier Reef extremely remote.

Now the *Navigator* was reported to be ready to leave, and to begin drilling in the middle of February. The first drill would be made five miles east of Mackay. If its test drills were satisfactory, drilling would begin in sections of the Barrier Reef forty to fifty miles offshore.

The Commonwealth Government, split as it was and still not committed to stopping the drilling, still made no real move to prevent it. It was clear that it was now unlikely to bring on the High Court case of its own motion, or it would have done so before.

With the results of the public opinion poll firmly under our belts, we began to think it might not be hard to raise from the public the money we would need to bring legal action against the government. We would probably, we thought, be able to bring this on the ground that Queensland's jurisdiction did not extend below low-water mark. This would force the High Court case that Sir Percy Spender had recommended the government to bring against a state. We might not win; but it was well worth a fling. And now that we had finished the main work of the poll counting – which had taken up so much of our time, also occupied with fighting applications for sand-mining leases in Cooloola – we might be able to muster the help needed for running such a giant appeal for the funds as would be necessary. We debated over the cost of full-page advertisements, over organising the writ, over how to handle the money as it came in.

The Commonwealth was, in fact, in a delicate position. It was clearly its duty – not that of private voluntary groups like ours – both to have the ownership of the territorial seas clarified and to protect the Reef from danger. But with its own internal conflict of interests – the Country Party was all for the mining interests, and it was widely suspected that much of the support for the Liberal-Country Party coalition came from them – and the lobbying strength of the great oil internationals, duty might well take second place, even to public opinion. The government was so weak that it might fall on other grounds than the territorial seas argument. Another election, when the last one had been so close, and when the anti-Vietnam protest was gaining strength every day, might put it out of office altogether.

But it would be highly embarrassing if small private groups like ours were seen to be having to take on the job of clarifying the whole offshore position – and if we advertised all over the country, on the scale we were proposing, that would be another count against the Commonwealth

Government. We could imagine the differing considerations that kept it more or less immobile.

The predictions of doom to the Reef were reinforced by the Great Barrier Reef Committee's continual pressure on the Crown of Thorns front. Endean, in December, said that 1,000 square miles of the Reef were now 'dead'. It would be a 'near-impossible job' to save it.

Endean contended that it would have been less difficult to save the Reef if groups of divers had been stationed on a few reefs in the Cooktown-Innisfail region, where it had first become serious, as soon as the danger had been reported. But the government had done nothing constructive, and had accepted unscientific advice (that of Ben Cropp and others) that the plague was insignificant and a recurring phenomenon, rather than that of qualified biologists. The only thing now was to try to stop the advance and save a few reefs selected for tourist value, by sending diving teams to collect the starfish by hand and to use underwater guns with formalin injections. 'When the full extent of the tragedy is known,' he said, 'Australia will be condemned by the scientific world.'

But meanwhile, with the Commonwealth refusing to move and the *Navigator* said to be almost on its way, there was a strange quiet from southern newspapers both on oil-drilling and on Reef ownership. We needed more publicity.

John had gone down to Melbourne towards the end of November, for what he hoped would be a final throat examination and for a great deal of dental treatment – he had lost a number of teeth in the various laryngoscopies he had gone through. He had been rather silent lately because of teeth extractions and illness. He was not coming back till late December.

I decided that I had better go southwards and try to get more publicity. If there was calm in the papers about the Repulse Bay project, there might be some reason for it. But we still had *The Australian* newspaper, with its then-strong commitment to conservation and its special interest in the Reef, to work for us. I rang Vin Serventy and asked if he could arrange a press conference and TV appearance.

I had the special advantage of being a kind of 'curiosity' showpiece in the conservation movement – a poet who spent most of her time on conservation was, after all, newsworthy. And though the problem of the Reef was far away from most Australians in other states and less immediate than it was to those in Queensland, it had to be brought home to them that they were losing, if the Reef was spoilt, something that was the responsibility of the whole of Australia, and of the world.

The trouble was, perhaps, that though most southerners did take an interest in the Reef's future, Gorton's statement had lulled many of them into believing that the drilling really would not take place, or that, if it did, the Commonwealth would take care that it did no damage. We had to dispel that idea. The *Navigator* was much too close for comfort, and to organise the writ would take time as well as the money we did not yet have. By that time, with oil-drilling already going on, it would be, as the Santa Barbara citizens had warned us, much harder to get the oil industry out than before the drills went down.

Vin organised everything most efficiently. I stayed with him and his delightful wife Carol; the reporters, the television crew, and a special columnist from the *Daily Telegraph* turned up, and I managed to put across a good deal of information that the Sydney press had either not known, or had forgotten. I was still not used to seeking this kind of publicity – and 'publicity-seeking' was one of the accusations oddly levelled by opposing interests against conservationists. But when the TV programme and the press articles came out, it was clear that the move had been a good one. All the southern newspapers began once more to carry the news that oil-drilling was about to begin in Reef waters and that the Commonwealth's assurances seemed to be coming to nothing after all. Again the Reef was a big news story.

John, in Melbourne, took advantage of this new surge of interest. He had an all-clear from the doctors, losing another couple of teeth in the process; his dental work was nearly done; and he had plenty of irons in the fire. He wrote from Melbourne in his old tone of confidence and energy, and he had much news. He was arranging for press interviews and radio and television programmes, but beyond that, he had been able to find out some things that both threw light on some of the difficulties we had met, and, as he said in his letter, 'made him properly and permanently hopping mad – a finely calculated, controlled rage'.

We had been more than puzzled by the opposition of the Reef's main tourist operators to the Save the Reef campaign, and their refusal to oppose oil-drilling, as well as their desire to play down the seriousness of the starfish plague. We had approached the Ansett company before, over the Ellison Reef case, for help, but without any luck. Now John had gone to see its head, Sir Reginald Ansett, whose chief island resorts were just north of Repulse Bay, and would certainly, according to a statement made by a retired harbourmaster who had made a study of currents and tides in that area, be among the first to be affected by any oil spills in the lovely bay just south of them. This, surely, John had thought, ought to concern Ansett. But Ansett refused an appointment to discuss this. The matter, he said, was much too controversial for him to be involved in. John set to work to find out what was behind this. He looked at the Shareholders' Register, and discovered that the principal shareholder in Ampol Transport Industries was Ampol, the second was Boral, the third W. R. Carpenter, and Ansett's own ownership of the shares was fourth on the list.

'Hence,' John wrote me, 'the confidence of the [state government] throughout the whole deal — they are backed by Ampol and Ansett in Repulse Bay. There would appear to be only one way out of it — the writ — but where to find the money I know not.'

But he also had in mind the Labor Opposition's commitment to stopping oil-drilling, and the good reception that the submission to the Trades Unions had had. He put his time in Melbourne to good account in that respect too.

And he had found another firm of solicitors in Melbourne and had put the problem of finding means of delaying, if not stopping, the drilling project to them. They had been most interested. Possibilities seemed to lie ahead. John publicised the proposal for issuing a writ in the Melbourne press.

'I am now optimistic we really can stop the bastards in Repulse Bay,' John wrote, 'but the issue of the writ is vital ... Ask the solicitors how much the issue of the writ would cost, and when should we serve it on the Queensland government to prevent the oil rig actually arriving in Australian waters?'

We were almost ready to go. The Christmas holidays had depleted our numbers a little, and we determined to move as soon as we felt the time was ripe, which would be very soon. John reported that the Commonwealth Government was anxious to act, but needed more time; the issue of the writ might give them that.

Meanwhile Theo Brown, of the World Life Research Institute, a research organisation based in California which was investigating tropical environment problems, and which was specially concerned by the state of the Reef and the Crown of Thorns problem, had run into a blank wall in his programme on the Crown of Thorns. He had begun this on Magnetic Island in November, after being granted permission by the State Department of Harbours and Marine. He was experimenting with sonic frequencies and their effect on the starfish; and now his experiments were suddenly prohibited by the Department which had originally issued the permission. Brown was thoroughly disgusted, and went on record as saying that he had grave fears for the Reef's survival. Apart from the very late adoption, in December, of the Endean recommendation to ban triton collection, the state government had done nothing either to control the plague or encourage further research. The Queensland government, he said, seemed to be making a deliberate effort to suppress the truth about the condition of the Reef.

The publicity this statement got set off another wave of accusations against the state, and the Commonwealth, over their inaction on the threats to the Reef. *The Australian* ran a full leader on Christmas Eve. 'Political dereliction', 'buck-passing', 'apathy', in a matter where public interest and concern had been so thoroughly demonstrated – it was, said the editorial, an 'insulting' response to that concern. 'When it suited them the responsible ministers were sympathetic; when they felt more secure they defended themselves by attacking their critics as "hysterical" or "anti-progress". The net result in terms of action has been negligible. The Barrier Reef,' it concluded, 'is a dramatic example of the environmental pollution questions facing public authorities throughout Australia. It must also be regarded as a touchstone of their intentions. The solutions are not beyond the wit and resources of the nation. All they demand is the will.'

The next couple of weeks were to see the biggest developments yet; to set all the state government's, and the oil companies', plans awry; and to vindicate the Prime Minister's stand, in spite of the fact that they originated with the Opposition and the trades unions. We were by now, through John's mediation, in touch with both the Prime Minister and the leader of the Opposition, keeping them informed of what we were doing and of the information we were getting. In spite of *The Australian's* strictures, we were sure that both were only too willing to step in on the Reef question, as soon as the opportunity arose. For once, we had leaders of both major parties on the conservation side; indeed, it had never been a party issue. The *Navigator* was nearly ready to leave for Australia after its long series of delays, to begin the mid-February drilling. All of us spent a very anxious Christmas.

More and more over the last decade, the pace of world-wide environmental decay had stepped up. We had done our homework on this, and we had examples of the trend near home that were in themselves bad enough. What worried us most was the obdurate attitude of politicians, industrialists, developers, even at times scientists who should have known better, to the whole situation. If even the Barrier Reef was the object of such determined attempts at vandalism, what hope had environmentalists anywhere of changing such attitudes?

We were despondent; all our efforts seemed to be coming to nothing. The Barrier Reef might be just one small part of the battle, but it was a crucial part, a symbol of humanity's failure to recognise its responsibility and of the whole relentless process of commercialisation and industrialisation, pollution, self-interest and political impotence. The 1970s were about to begin, and they promised to be far more disastrous than the 1960s had been. And still, even in the conservation camp, there was caution, temporising, playing for advantage, and attempts to come to terms with, rather than face, opponents whose enormous power discouraged many. If it had not been for the public backing for protection of the Reef that we knew existed, we might have given up hope.

As it was, we were going to get that writ issued before the *Navigator* arrived, if it beggared the lot of us. And, determining that it was long past time that politicians were really made to face up to the situation they were creating overall, I, as president of the society, wrote an open letter to the Premier and Prime Minister, and sent copies to all the newspapers. I said that the beginning of the 1970s faced all governments with the need to take swift action on the problems of pollution and the conservation of the environment, but that so far none appeared to be willing to face the crisis. If governments did not act in the face of the rising public concern, it would be for them to take the blame for the results.

All the newspapers to which the letter was sent printed it; only the Brisbane *Courier-Mail* made certain cuts. *The Australian* used it, and an interview as well, to usher in the New Year.

We also prepared a statement on the society's attitude on the Great Barrier Reef:

The Executive of the Wildlife Preservation Society of Queensland reaffirms its belief in absolute protection of the Great Barrier Reef against any form of exploitation which involves or which may initiate unnatural and massive disturbance of the marine ecosystem of which the tropical coral reefs are an integral part.

Specifically, we believe that there should be:

(1) absolute protection, *sine die*, against mining for limestone or other minerals, and against exploration for and extraction of oil or natural gas. Recent experience at Santa Barbara and elsewhere shows that it is not humanly possible to prevent or control massive oil pollution from drill-holes, wells, pipe lines and tankers: 'accidents' are inevitable.

(2) progressive and controlled development of tourist and fishing

industries based on the results of adequate and continuing scientific research to ensure that the type, location and extent of developments do not damage the Reef resources in the long or short term. Our attitude towards the proposed development of a marine oil industry on or near The Great Barrier Reef is therefore unequivocal and our policy of preservation (as opposed to 'conservation' meaning 'controlled exploitation') would be maintained until it were shown that no alternative sources of oil existed elsewhere.

We were going to use the first issue of *Wildlife* in March as a special Barrier Reef issue, to tell people what they were losing, and to keep the agitation in the public mind, even if the drilling had begun. I was to leave for India, for a cultural conference, early in the New Year.

Our hopes for a final decision on a union ban were rising. But it seemed too much to expect. We would not know for sure whether it would be implemented until the second week in January. And if it were, it would be the first time, not only in Australian history but as far as we knew in world history, when the trades unions had taken a step that went so far outside their traditional boundaries of interest. If it were only a threat, it would not be enough. And the question of how it would be carried out, and how firm the unions would remain, was still uncertain.

But here we had another unexpected stroke of luck, as far as publicity went. There were others who were looking at the coming decade with dismay; environmental workers in the United States were putting great pressure on the President, and environment groups were growing stronger and stronger. On 3 January the headlines were suddenly full of the very subject on which we had written our open letter to our own politicians. President Nixon, signing a bill to set up an advisory council on pollution, had unwittingly struck a blow for the Great Barrier Reef as well. Parts of the United States, he said, would be unfit to live in within ten years unless Americans acted now against pollution of air and water. 'It is literally now or never,' he said. 'The 1970s absolutely must be the years when America pays its debts to the past by reclaiming the purity of its air, its waters and our living environment.'

The newspapers made this a front-page headline everywhere, and ran articles on Australia's own problems of smog, and pollution of water and air. A 10,000 million dollar project to clean American waters had been proposed; the implications for Australia were clear enough. 'Pollution has supplanted Vietnam as the big worry in America,' *The Australian* wrote '... It may be too late for America ... it is not too late for Australia, but we must act now, for 1970 could be our last chance.'

All this publicity was another push on the way for the trades unions. The time was right; the *Navigator* was soon to set off from that polluted country to add to our own problems.

On 6 January the newspapers had a new set of environmental headlines. UNIONS LIKELY TO BAN WORK ON COAST DRILL, said the *Courier-Mail*. The Transport Workers' Union had decided to recommend to the Queensland Trades and Labour Council that it convene a meeting of all the unions affiliated in Queensland to consider imposing a total ban on drilling.

Senator Georges, president of the Save the Reef Committee, made public the telegrams he had just sent to the owners of the *Navigator*, to Ampol Exploration, to Ampol's chairman of directors, and to Japex. The failure of state and federal governments to ban drilling, in the face of the strength of public opinion against it, meant, he said, that direct action had to be taken. 'Therefore, before the *Navigator* leaves for Australia, I warn those in control that I intend to launch a campaign to declare the vessel black and to withhold services of labour and essential goods for its operation.'

The work John had done in preparing the original submission to the ACTU conference, and again in Melbourne during the early part of December, was triumphantly vindicated. He had never lost a chance to talk to politicians, on either side; both Senator Georges and Senator Keeffe, as Queensland senators, had listened to what he had to say, as well as the Prime Minister, the leader of the Opposition, Gough Whitlam, and others. Now at last, as he gleefully wrote me, the breakthrough had really come. 'It has taken two-and-a-half years to find the weapon. This is it.'

The Australian, printing the reply by Ampol's chairman to the Senator's telegram, made a rather fuller report than some other sections of the press. It read in part: 'While understanding your obvious anxiety to protect the Barrier Reef, may I assure you that Ampol, which incidentally is an Australian company, is equally as vitally concerned with the preservation and conservation of our national heritage, and always has been. It surprises us that you have not previously approached us directly to discuss the scientific facts of the operation (*sic.*).' But what followed was no more than a reiteration of what the Minister for Mines had already said about the 'stringent safety precautions' and the stationing of an inspector aboard. There was going to be no question of giving in at this stage.

The question of whether Ampol was, or was not, in fact an 'Australian' company was one we had looked at before. Australian-registered it might be; its actual ownership might be a different matter, we had ascertained. Senator Georges, in his telegram, had suggested that there might be a campaign to question this. Since Ampol's main advertising laid stress on its Australian character, it could be an important point – it was one we had decided to put aside for further inspection as a possible weapon.

The Australian's report went on to point out that the Repulse Bay rig was only the first of many scheduled to begin operations in Reef waters. One of the earliest operations, we expected, would be in Princess Charlotte Bay, on the Exoil-Transoil lease officially known as 8P. An airstrip had recently been built on Lizard Island, in the neighbourhood of the lease area, in the far north of Cape York. It was supposed to be for tourist purposes. But Princess Charlotte Bay was almost without settlement. Even the fishing industry had not yet established itself there; the great stands of mangroves that lined the shores of the Bay were most important fish-breeding areas, but there was little on the mainland to indicate that it would yet make a feasible tourist area.

We were all exultant over the prospect of the ban. We stayed our hands on preparing the writ; it might not now be needed.

The *Navigator* now was said to be leaving; it would take about forty-five days to reach the Queensland coast. Representatives of both Japex and Ampol were still firm, outwardly at least, that the drilling programme would go on.

However, the Minister for Mines said that there was nothing the state government could do about a trades union black ban. 'If the trade unions want to deprive the people of Mackay of the money that these drillers would pay for their supplies, they can go ahead. I am sure they will be able to get them elsewhere, or bring their supplies with them … These people are simply standing in the way of progress, and they want Queensland to be the only state without offshore exploration.' He also revealed that the drilling conditions were so stringent that 'two companies who were interested in offshore leases refused to go any further when they saw the conditions we have imposed.' He would not say what they were.

On 7 January, the Trades and Labour Council said that because of the magnitude of the problem, a special meeting of all affiliated unions and the council's disputes committee would be called to discuss the Transport Workers' proposal. It might not be possible to hold this meeting immediately. It added that, if it were necessary to call on the ACTU inter-state executive to help in implementing the ban, it was certain that other state representatives would back the Queensland move.

The state government was now thoroughly alarmed. The Mines Minister fell back on saying that Mackay could lose thousands of dollars because of the ban, in sales of food and supplies. Mackay was in his own electorate. Because of this, our Mackay branch had come under a lot of pressure and met many problems in its opposition to drilling in Repulse Bay. It had decided not to conduct the public opinion poll, but it had circulated a petition instead. This got 1,304 signatures, in an area which was delicate indeed. But Mackay residents, whatever their political persuasions, had a most beautiful area to live in. Repulse Bay, north of the city, is a great stretch of blue water, with a rugged rainforest-clad northern coastline, and the islands – those famous tourist resorts – offshore. Many Mackay people were more than doubtful of the benefits of oil-drilling that might spoil those lovely shores and islands. The petition had asked that all permits for prospecting and test drilling for oil on the Great Barrier Reef, so far granted or pending, should be cancelled. The TLC secretary quoted this in reply to Mr Camm.

Japex was in at least as awkward a position as Ampol. It was fifty per cent owned by the Japanese government; and with the Prime Minister's attitude on drilling clearly stated, Japex (Australia) was unhappy. The question of offshore ownership was only one of the difficulties it might face. Japex and Ampol held a conference on 7 January to discuss the whole position. The fact of the Japanese government shareholding was delicate. But 'this well is considered to be purely a private business deal between ourselves and Ampol', said the Japex representative. Under their agreement with Ampol, the well had to be drilled by 30 June. If the black ban could stave off the drilling even until then, we thought, something would be gained; and the Commonwealth might have time, behind the scenes, to negotiate some kind of agreement with the oil companies, and launch the High Court case which we ourselves would have had to bring.

The newspapers seemed to be all in favour of the black ban, though they blamed the two governments for making it necessary. 'The black ban,' said *The Australian*, 'will have an unprecedented measure of public support and will probably succeed. It deserves to.' It pointed out that there was nothing illegal, so far, in the ban; unless an industrial tribunal were later to rule against it, it could not be stopped by any other means. If indeed there were a tribunal ruling against it, 'the responsibility will lie directly with the two governments and particularly with the Queensland government.'

Turning to the question of compensation which the state government insisted would have to be paid to oil companies if the leases were not available to them, *The Australian* said: 'If compensation is necessary, the government is entitled to assume that the Barrier Reef campaigners accept that fact, although it will be accounted as the price of a government mistake.' And it went on to quote Nixon's speech, and Russell Train's words to a recent oil industry conference on pollution in the United States: 'The people of the United States are insisting on a reordering of national priorities, and we ignore or miscalculate the significance of this movement at our peril. I predict that the quality of the environment will become the major issue of our time.'

For once, it seemed, Australia was not only up there with the best of them, but even a jump ahead. No trades union in the US, that we knew of, had considered taking this kind of action. But then, perhaps, they had not had as much, of such obvious value, at clear stake. To keep oil out of the Reef was an issue that no one – except the Queensland government, it seemed – could dispute.

I left for India with the result of the trades union meeting still unknown. I had a habit, it seemed, of being overseas for the big news.

8

A Halt, and a Surrender

In India I was quite unable to forget what was going on in Queensland. Would the unions really go ahead? Or would we have to issue that writ? I looked at India with our own battles occupying half my mind.

India was seen by my Australian friends as a country with 'insoluble problems' – as indeed it was. But it seemed to me that Australia's own problems were going to be as bad, very soon, though differently so.

Australia, unlike India, had produced no religion, no philosophy, little art of its own. Its brief history was a rage of purely material exploitation; its treatment of its Aboriginal precursors was at least as bad as India's treatment of the lowest of its outcasts. We thought ourselves rational, educated, enlightened; we had the benefit of almost every advantage of the twentieth century. Yet we looked likely to destroy our own country in far less time than Indian cultures had taken to reach their own point of poverty, land exhaustion and over-population. And in doing so, we would have contributed far less to the world than India had done.

On 8 January there had been replies to Senator Georges' telegrams. The drilling contractors for Japex (Australia) Pty Ltd had answered that drillings proposed were 'tests only', and would take thirty to forty-five days to complete. 'At the conclusion of this operation we know of no further plans for drilling in the area,' they said. They added that the

Navigator would have the most stringent anti-pollution precautionary equipment ever fitted to an offshore rig – not even 'so much as a cigarette packet' would be allowed to be thrown overboard. They were convinced that 'no danger whatsoever exists to the Great Barrier Reef'. But they added ominously that they had plans to 'expend here further large amounts' – on what, they did not disclose.

They suggested a meeting with Senator Georges. He answered that he would meet them any day the following week. 'I need to be convinced,' he said, 'that your drilling operations will not lead to further extensive drilling of the Great Barrier Reef. I believe that your drilling will create a precedent and will increase sharply the demand by other permit holders, thus magnifying the risk of a blowout.' It would also lead to more mineral exploitation of the Reef before laws were enacted and research undertaken to protect it from the results.

He also answered the telegram sent by Ampol Exploration Ltd, with its simple reiteration of assurances that few people any longer believed. 'Can your company assure me that having completed drilling in Repulse Bay with permit 162P, it will not proceed with drilling on Q12P which it controls and which extends over 4,000 square miles of water in the reef area?' he asked. And he also sent a telegram to the constructors of the rig, Zapata Australia Pty Ltd, asking whether the company was prepared to pay compensation for any damage caused to nearby tourist resorts and to Queensland people by the rig's operation.

The feeling against drilling, and against the authorities proposing it, was kept at fever pitch by the daily reporting of moves and counter-moves, challenges and assurances. The *Sydney Morning Herald* featured that Saturday a drawing by Charles Blackman with a fairly inferior and hurriedly-produced verse of my own – I was not in the mood for verse, even if I had had time to work on it – to go with it. Letters were published from people all over the country protesting against the drilling.

It was a concerted and spontaneous outcry; and it (or some other factor) worked. Ampol's chairman announced on Tuesday 13 January, that the drilling would be postponed, and offered $5,000 to help pay for an inquiry into the effect of oil drilling on the Reef. He had, he said, sent telegrams that day to the Prime Minister and the Premier of Queensland asking for a joint committee to examine the problem – and also the Crown of Thorns problem. He added that 'during the past ten years six holes have been drilled by other companies in the reef area, including three since 1968, without arousing any public outcry or concern'. (These, which included the Swain Reef drillings, had been authorised by the Queensland government and had, we knew, resulted in much pollution. The fact was that information on these drillings was very hard to get and was not publicised – and it had taken a long time for us to work up to the point where we could actually count on public support for stopping the drillings.)

On 14 January the Premier said he had had no official communication on the proposals by Ampol to Japex to delay drilling. He was on the Gold Coast, where a by-election was later to be held, and he would confer with the Mines Minister next day on the proposal. He was concerned, he said, that 'supposedly responsible people' could bring about the present situation; he was definitely opposed to Ampol's idea that an inquiry into the Crown of Thorns plague should be held at the same time.

The Transport Workers' Union State secretary in Queensland said that Ampol's interest in safeguarding the Reef was 'belated'; until now, it had not shown any interest in preserving the Reef; and a committee of inquiry was probably going to be ineffective: 'committees have a habit of being hand-picked'.

The Mines Minister's reaction was that Ampol's proposal was 'a pity'. 'Most other countries in the world capitalise on offshore oil and gas. Every other state in Australia is encouraging offshore installations … However, if the company concerned declines to go ahead with the project, that is its responsibility. I realise it could not continue with this threat emanating from Senator Georges and the unions.'

Des Connell, of the Littoral Society, said it would be a most appropriate time for the Commonwealth Government to step forward and take some decisive action on the whole issue of preservation of the Reef.

The general manager of Japex (Australia) said it would be 'a few days' before a decision could be made on stopping the drilling; he was telephoning Tokyo that day.

On 15 January it was reported that the Queensland government wanted the drilling in Repulse Bay to 'go ahead immediately'. Obviously the Mines Department and the Premier were thoroughly upset that Ampol had taken action on its own account – 'disappointed', was the Premier's word. 'The information to be gained from drilling a well in this area would be of immense value in assisting the assessment of oil-bearing possibilities in the waters off Queensland.' He had 'received advice from recognised world authorities'. To this Senator Georges answered that if the drilling proposal went ahead, the black ban arrangements would be brought on and extended. 'I challenge the

Premier to name his experts and publish their advice. Has he sought the advice of a top biologist or ecologist, or has he depended entirely on the advice of his mines inspector?'

The Premier did not accept the challenge.

The Australian, in its leader that day, said that Ampol's offer of an inquiry and suspension of drilling was 'certainly commendable; but it suggests two things: that Ampol's carefully-nurtured image as the "Australian" company is in danger if it pursues a course inimical to public feeling, and that there is a possibility of reef pollution – an admission that obviously embarrasses Mr Bjelke-Petersen'. It further pointed out that the *Navigator* was now ready for Japex to lease, and that there was a serious shortage of oil rigs in the world oil industry. 'Nobody concerned is going to let this one remain idle. How meaningful then is Ampol's "recommendation" to Japex? What are their contractual obligations to each other? In these circumstances, it is to be hoped that Mr Gorton will do more than congratulate Ampol for its concern over the reef.'

Mr Gorton *had* congratulated Ampol on its plan to defer oil-drilling. It was a 'nationally responsible attitude'. But Mr Bjelke-Petersen thought differently; he said that he 'rejected' the proposal for a Commonwealth-state committee. If an inquiry was going to be held, moreover, the state government would also reject any payment towards its cost by Ampol.

The Academy of Science in Canberra, which had long been waiting on the sidelines for some decision on its hoped-for scientific survey, had got in touch with the Premier and offered to conduct this survey. The Premier rejected this, too. 'There could be no purpose to a further investigation,' he said, 'an expert survey already has been completed in a most competent way.' (What this survey was, unless it was the Ladd Report, was not made clear.)

Mr Gorton had said that he believed the Barrier Reef was of such importance to Australia and to world science that it was 'essential there should not be the slightest chance of any damage occurring from such drilling'. But one problem, which had not so far come to public notice, was now canvassed. The proposed drill in Repulse Bay was certainly in Queensland waters. There was doubt about the states' jurisdiction below low-water mark, but it did apparently have jurisdiction over waters within a line drawn from headland to headland of enclosed bays not over twenty-four miles wide – and Repulse Bay was one.

This caused government legal advisers to question whether the Commonwealth Government could actually institute an inquiry on its own account. Did the Commonwealth have any jurisdiction there? The prospecting permit had been issued under joint Commonwealth-state regulations; it did cover some of the Ampol area, but possibly it did not cover this particular drilling project. Queensland, it seemed, would have to agree before any inquiry was held – and perhaps, under the joint Commonwealth-States Offshore Petroleum Resources agreement, no inquiry could take place at all without Queensland's consent.

Now the battleground was complicated by yet more arguments. The fight had drawn in more and more protagonists: geologists, biologists, governments, oil companies, trade unions, even nations (since no reply had yet come from Japan) were involved in it; and now legal advisers had joined in. Conservationists were beginning to stare in amazement at the complications they had brought on. Two simple facts remained: the need to protect the Reef itself, and the proven strength of public opinion.

Mr Gorton could only press the state government to agree to the joint inquiry. He did this on 16 January. He offered to appoint experts from Australia and overseas for a public investigation of the dangers of oil-drilling. Queensland replied that, provided the proposed drilling at Repulse Bay was not included in the inquiry, it would agree to accept the offer. 'Sufficient inquiries into the area have already been made,' said the Premier, repeating that he wanted the drill to go down in Repulse Bay immediately.

Mr Gorton released the text of a telegram to Ampol Exploration. In it he had said that 'in my view the slightest danger is too much danger'. And, he added, 'it is my hope that other companies which hold prospecting leases will follow your lead pending a complete investigation and report … As you know, the Commonwealth Government was bound to confirm the oil leases issued by the state on and near the reef and is virtually unable to prevent drilling on these leases if the State agrees to drilling.'

He also rejected Ampol's offer of money towards the cost of an inquiry. 'I believe you would agree it might be better if an investigation was in no way financed by an organisation connected with oil,' he said; and he also rejected the idea that the investigation should cover the Crown of Thorns problem as well as oil-drilling.

It was now that the lack of that High Court case, suggested by Sir Percy Spender many months before, became an embarrassment to the Gorton government. If the government had had the courage to take up that challenge, it would already probably have had the weapon it needed to stop the drilling and withdraw the leases already issued. But in the face of the powerful state governments, frightened that the Offshore Oil

Agreement might be proved invalid, it had decided to take the position that the Constitution prevented overriding the state governments. Unless it could show that drilling was actually affecting some accepted Commonwealth responsibility – fisheries for instance – it therefore had no jurisdiction.

Whitlam, however, had taken the opposite view, saying the Labor Party was convinced that the Commonwealth already had the constitutional power it needed, and that if the ALP came into power it would stop the drilling. But it was not in power. The state government fell back on the request it had made many months before – which was now to be fully revealed – for an advisory committee on Reef resources.

The Premier telegraphed Mr Gorton: 'As far as my government is concerned, a basically similar but wider proposition regarding drilling on the Barrier Reef was put before you by my predecessor (the former Premier, Mr Pizzey) in a letter of 7 June 1968, to which a detailed reply is still awaited. The proposals then made and the supporting information given were, and still are, pertinent to the question of Reef drilling. I see no reason why they should not appropriately provide the basis for any Commonwealth-state investigation now. Consequently you will appreciate that until such time as you furnish me with your Government's views on our proposal ... or any alternative you care to advance, I would prefer to defer a decision on your telegram.'

The proposal referred to, made at about the time when the Ellison Reef decision was finally taken by Mr Camm, had been addressed to the then Prime Minister, Harold Holt. Evidently it had been the one that the Queensland government representative had spoken of at the ACF symposium.

Mr Gorton told reporters that the Queensland government had then proposed to set up a committee to advise on the whole question of control and exploitation of Barrier Reef resources – fishing, tourism, scientific research and mining. This was to have been an eleven-member committee, and the Queensland government had suggested that the Commonwealth might nominate one member.

'This suggestion,' said Mr Gorton, 'was entirely different from the present proposal, which is that drilling for oil be halted until a joint inquiry into possible dangers of drilling has been held, and has reported publicly to both governments ... This is the question which now awaits the decision of the Queensland government.'

The Commonwealth's attitude on the state government's 1968 proposal had been that it could see little point in having only one member on a committee which would advise only the Queensland government, and it had never formally replied to the request. Such a committee was the Queensland government's business, and it had not yet been set up.

The *Courier-Mail*, printing the usual letters from more or less outraged members of the public about the Reef drilling proposal, commented in a leader that the inquiry proposed in 1968 had not specifically mentioned oil-drilling at all.

The legal argument now seemed to be that the Commonwealth could not block leases for oil-drilling that had already been issued – only future ones. Only the Queensland government could act to block the present leases.

'Mr Gorton, who is treading the conservation path with praiseworthy single-mindedness of purpose,' the newspaper said, had only one weapon to use, and that was public opinion. But this could be enough to carry the day, even in the face of Queensland government opposition. More and more organised protest was emerging. A Housewives' Reef Preservation Committee had been formed, and had sent letters to the Premier, the Mines Minister, and the Minister for Tourism, challenging them to say whether they would resign immediately if a blowout occurred in Reef waters. Every paper carried columns of letters. The state branch of the railwaymen's union, the Australian Federated Union of Locomotive Enginemen, said it was going to support the move for a black ban. The meeting of all Queensland affiliated trades unions was to be held in a couple of days to discuss the recommendation, and the AFULE was an influential member.

The Premier said that the state government would agree to a drilling suspension, provided the companies concerned made no financial claims on the government. In Tokyo, Japex officials said they were still waiting for a decision from their Australian representatives.

The Queensland Liberal leader, Mr Chalk, now said that in the past he had taken the attitude that, because the government had made undertakings to the oil companies and they were now deeply involved in the drilling proposals, there was no basis for repudiation. But now that Ampol had approached Japex, the situation was different, and there were 'points to be discussed' with the Country Party cabinet members and the Premier. Evidently the state government coalition was on the verge of a split, if the Country Party stayed obstinate.

The legal argument that the Commonwealth had no power to revoke leases it had already approved, was challenged by the Federal Labor Party. Dr Patterson declared that a future ALP government would repudiate the joint Commonwealth-states offshore oil agreement and would see that only the Commonwealth should determine the development of offshore resources. Sir Percy Spender told the press that he still held to the view that dominion over the seabed and the natural resources of the whole of the continental shelf from low-water mark belonged to the Commonwealth. This supported the ALP's contention that the Commonwealth could stop the drilling of its own power. Sir Percy urged that the High Court case should be brought on for the sake of the present and succeeding governments.

On 17 January, the Premier reverted to the theme that the Commonwealth had still not replied to the 1968 request from Queensland for Commonwealth participation in a committee on the natural resources of the Reef. He reiterated that the Repulse Bay drilling was completely divorced from any investigation into oil-drilling on the Barrier Reef, and that the question of drilling on the Reef had been considered from every angle – not only by the Mines Minister but by 'every member of Cabinet' – and was based on the 'unbiased and unprejudiced recommendations of both Australian and overseas experts'.

But that night he finally gave in, and consented to the Gorton proposal for a joint Commonwealth-state inquiry. He would recommend this to the cabinet meeting on the following Monday.

But, as the press pointed out, he made no secret of his reluctance to agree to either the halting of drilling or the inquiry, and gave no indication that he would accept the findings of the inquiry if it went against drilling. Every statement made by the Queensland government indicated that it felt there was some acceptable minimum risk of damage to the Reef, while public opinion could not do with anything less than ironclad guarantees. And, said *The Australian*, the Repulse Bay drilling should not be excluded from the inquiry's terms – 'there are deep questions of aesthetics involved in the area'.

Now that the inquiry was really to be held, we on the conservation side were starting to look anxiously at our position and its responsibilities, and at the very small resources we had to meet them. We had no money to bring scientists from overseas to give evidence, and since few, if any, then in Australia had experience of the effects of oil pollution on marine life, such scientists would be crucial witnesses. If we had to present their written submissions, these might not be acceptable to the inquiry without personal cross-examination of the witnesses themselves. The oil industry would have no such problems. The president of the Australian Petroleum Exploration Association announced that APEA would 'provide as much expert evidence to the inquiry as they could muster', and their 'conservation committee' was meeting immediately to discuss what to do.

Australia's long starvation of marine research meant that there were painfully few top marine biologists who had qualifications in marine ecology, and most of the important work done in this regard had been done by Fulbright research scholars and visiting overseas biologists, most of whom were now based in their own countries again. Even to pay their fares was far out of our reach, let alone to reimburse them for their time and trouble. We asked the Premier whether any money would be available for the expenses of witnesses. No, he replied; all witnesses would have to pay their own expenses.

Our David-and-Goliath problem looked worse than ever. We would be facing the giant oil interests without a single sling-stone.

The Monday cabinet meeting seemed to have run into difficulties. On 20 January, the Prime Minister sent a hurry-up telegram to the Premier. Would he officially endorse the proposal for the inquiry, so that discussions on the composition of the committee could begin? The proposal was for a 'truly joint Commonwealth-state inquiry which will report to both governments and which will make public its reports'. It was a hook the state government did not want to swallow.

The Commonwealth had now closed down all the drills in the Gulf of Papua which were under its control. This answered the Premier's accusation that the Commonwealth itself was engaged in drilling in Reef waters.

Finally, on 21 January, the Premier said that the state would only accept the inquiry proposal on terms to be decided at a special conference. He would take the Ministers for Mines and for Justice to Canberra for discussions, but they would have firm ideas both on the composition of the committee of inquiry and on its terms of reference.

Many people had been surprised that the Prime Minister had not intervened long before, and some had doubted his good faith. His real problem, of course, had been the credo of his party and of its coalition partner, that the states should be allowed to go their own way with as little interference as possible. But on the other hand, the Commonwealth had been equally reluctant to cede its potential offshore powers to the states. This was why there had never been an official answer to the Queensland government's request for Commonwealth participation in the advisory committee on the Reef's future − a committee which would have had only one Commonwealth representative and would have reported to the state. To accept this would have been, in effect, to admit that Queensland owned the Barrier Reef, and no doubt other states would have taken advantage of this.

However, the Commonwealth Government, though not its Prime

Minister, wanted to keep the offshore question under wraps. Perhaps Queensland would be amenable to private negotiations and matters would not have to come to an open showdown. But the La Macchia case, and Sir Percy Spender's open support for the principle that State ownership ended at low-water mark, had forced its hand.

John knew Mr Gorton's situation well. Since the press was questioning his goodwill, John had decided to see that justice was done. On his way north from Melbourne, he stopped in Sydney to do a little background briefing.

The Australian now ran an article based on this. It revealed that the two judges in the La Macchia case, Sir Garfield Barwick and Sir Owen Windeyer, had been specifically asked not to give a judgement on jurisdiction below low-water mark, but they had felt such a judgement was legally necessary. This had set off a train of consequences, among which was the fact that the Opposition was able to exploit it and to embarrass the Government particularly over the hot public question of the Barrier Reef.

The coalition parties had to recognise that this was a powerful election point. While they themselves had a strong internal lobby on the side of oil-drilling, the October election made this lobby unpopular. Drilling on the Reef could clearly swing public opinion, and perhaps not in Queensland alone. Labor was making all the running on this, and getting all the good publicity; the trades unions were making the popular moves. But the coalition still refused to move, and Mr Gorton's hands were tied.

'About the only thing that could have saved Mr Gorton at this stage,' said the article, 'was for Ampol itself to come in on his side, and against Mr Bjelke-Petersen, and by what may or may not have been a complete coincidence, this was exactly what happened.'

To set up a public inquiry would give the Prime Minister time and a chance to attack from another direction. If the inquiry showed that drilling the Reef would have an effect on a constitutional responsibility of the Commonwealth — such for instance as fisheries — the Commonwealth could intervene in any case and the Liberal Party could hardly object.

As the article said, 'Mr Gorton has done well to find an opportunity where it is not merely politically expedient, but politically possible, for him to move. And even Mr Gorton's worst enemy could not deny his. genuine concern — and love — for the Reef.'

This cleared up any doubts about the Prime Minister's good faith, and brought into the open the Liberal and Country Party's own part in the whole proceedings.

Now all depended on how much influence the Prime Minister could bring on the inquiry's terms of reference, and on the actual appointments to the committee. Both these would be crucial. If another 'hand-picked' committee was set up, and if the terms were too narrow, the inquiry could be hamstrung. We knew that Queensland's ministers, if they agreed to the inquiry being held at all, would certainly want a quick and easy justification of their own policies.

But the Commonwealth had bargaining power. The Queensland government wanted other things beyond oil-drilling on the Reef. One was a new power-station for Gladstone, where Comalco and other enterprises were demanding more electricity for their industries, and which needed Federal moneys. Perhaps a deal could be done.

Such a deal might be expensive, but even more costly would be the kind of inquiry we wanted – a thorough and lengthy one, with numerous overseas witnesses not brought at our expense. If we were to face this inquiry at all – and it was not at our request that it had been brought on – it could only be useful if it were impartial and wide-ranging, and took as much time as it would need. And we felt an inquiry held in Australia alone would not be enough. We wanted not only overseas evidence, but overseas inspections too.

As we waited to hear the outcome, the Queensland Trades and Labour Council held its all-important meeting. On 21 January, it approved the application of a total black-ban by all affiliated unions on oil-drilling on the Great Barrier Reef. A committee of three representatives of the Federal Parliamentary Labor Party and three from the State Labor Party was to decide how the ban was to be applied.

We were saved by the bell. This confirmation seemed to make it most unlikely that the oil companies would or could make any further moves on their Barrier Reef permits. We would not, after all, have to issue the writ.

9

Quarrels and a Victory

The union black ban on the Mackay drilling was spectacular and unprecedented. It made headlines and front pages throughout Australia. Some argued that it was not for the unions to take action which overrode government decisions, and that the precedent was dangerous.

But probably no trade union decision anywhere in the world had been more generally popular. The Trades Hall in Brisbane reported that not one of the letters and phone calls that poured in after the announcement had been anything but congratulatory.

It remained now to be seen what the reaction of the state and Commonwealth governments, and of the oil companies, would be. The split between Mr Gorton's views, and those of many in the Federal coalition parties, was well known. But the clear public support for stopping the drilling strengthened his hand. Would the six companies which held drilling permits in the Reef area agree to suspend operations during the public inquiry?

It was now announced that all but one of them had decided to do so. But they said that to hold their hands for more than six months would be unacceptable. They had already gone a long way with preliminary surveys. Naturally they wanted quick action and a brief and limited inquiry.

The one company which had not yet decided was Japex. Ampol's suggestion that the Repulse Bay drilling be suspended had put it in an awkward front-line position, for it had already assembled equipment near Mackay and spent money on its rig, and the *Navigator* was said to be approaching its destination. Reports said that Japex might pull out of oil search in Reef waters altogether, because of the expenses it would have to meet in waiting for the inquiry to end.

The members of the proposed inquiry had not been chosen, nor its terms of reference announced. The inquiry might not start for months, and no term had been set for its final report to be presented.

On 21 January, Japex announced that the company would agree to suspend its drilling, if Ampol reimbursed them for 'all reasonable expenditure'. In a telegram to Ampol, the company said that it was surprised that Ampol had delayed expressing concern 'until this time'.

Ampol replied that the expenses were clearly Japex's risk, and the decision on drilling was up to Japex. For that reason, it would not reimburse the expenses, but it would defer any contractual obligations without penalty. Ampol also denied that it had not expressed its concern earlier. The managing director said that a phone call had been made to Japex on 6 January and a telegram had been sent on the thirteenth. They released the text of the telegram. It read, in part:

In view of grave fears expressed by sections of the community that the drilling constitutes a threat to the Great Barrier Reef, it is our opinion now that industrial action will prevent the drilling of this offshore well by you, when the rig arrives on site. As you would then be involved in heavy expenses relating to rig mobilisation fees and standby charges and other expenditure we believe that it would be in your best interests to postpone the drilling. We are recommending to the Federal and Queensland governments that a properly constituted committee should be set up.

Japex now referred Ampol's refusal of compensation to Tokyo for decision.

On 25 January, the press reported that the Repulse Bay project was likely to be abandoned. Only the question of an arrangement to cover Japex's costs now stood in the way of a final decision. Japex was claiming to have spent at least a million dollars, and its costs would rise with every day of delay. If the *Navigator* reached the drilling site, another $340,000 would be incurred.

Where in fact was the *Navigator*? It had several times been reported to be already on its way, but it now appeared that it had only just left Texas.

The article in the *Sunday Mail* which revealed this, contained other information of interest to us. It went on:

Some experts said yesterday that a decision to abandon [the drilling] would be a victory not for public opinion but for oil business propaganda. They said that behind the scenes much of the opposition to the Repulse Bay project was engineered by American oil interests, and said that American oil companies had been involved in previous Australian offshore ventures. The interests are said to have opposed the entry of a Japanese company into the area for fear this would upset their projected price structure.

(Australians have been warned to expect a lift in the price of petrol from three to five cents a gallon in the use of offshore oil.)

This is why so much fuss was made about the Repulse Bay project while six previous drills on the Reef attracted no attention, said one man.

What truth there was in the unnamed 'experts'' revelation about the attitude of American oil companies to Japex we had no way of knowing. We had no interest in the internecine squabbles of oil companies, and certainly no communication with them. Our shoe-string operation had been financed wholly from our own private funds and the small donations made by our own members and other people concerned for the future of the Reef. Most of these had come in after the Santa Barbara disaster. Since our organisations did not have the advantage of being eligible for tax-deductible donations – except for the Australian Conservation Foundation, which had made very little contribution to the battle, financially or otherwise – these had indeed been small, amounting to a few dollars here and there.

As for us, we were opposed to drilling by any company on the Reef, and had said so publicly and firmly. Much the same now applied to the public, too – it would not now have mattered where a drill went down or who put it there, as far as most people were concerned. Indeed, very likely an American drilling would have aroused even more opposition, after the demonstration at Santa Barbara of the American capacity to prevent blowouts.

Our withers were therefore unwrung. But we reflected that, since exactly the same organisations were being accused of being tools of anti-American interests, in the fight against American-owned companies exploiting the Cooloola sands, it was a fair demonstration of the impartiality of those organisations that we should now be represented as tools of American interests themselves. Either way, we got no money for the job!

I came back after my fortnight in India, and was plunged at once into conferences on our prospects and future tactics. It was a dizzying time. With the prospect of a public inquiry, we had much to think about, and the oil industry was already organising its forces.

On 28 January, the Chairman of APEA released a statement which was, he said, to be the theme of his address to the APEA's annual conference in March. This was, significantly, to be held in Queensland, on the Gold Coast.

The world, he said, would require as much petroleum in the next ten years as it had consumed in the past 110 years. This would 'test to the full the capabilities of the petroleum industry. If the oil industry is to remain the productive and responsible component of society that it is today, an equilibrium of attitudes will have to be maintained. There must be some degree of balance in what society wants from the oil industry, and what licence the industry must have in order to supply these requirements.'

Evidently, the Barrier Reef controversy was to be a main theme of the conference of petroleum explorers. We fully agreed that there must be a balance. So far, it seemed that there was nowhere in the world where the oil industry was not permitted to drill. In the United States, environmentalist action against the proposed Alaska pipeline was being hotly opposed by oil interests. The requests by Californian citizens to keep oil out of the offshore areas had been ignored before the disaster. Now the Great Barrier Reef was another arena for the argument.

This statement was evidently timed to influence public thinking towards oil-drilling on the Reef at a crucial time. The meeting of the Commonwealth and state governments to discuss the proposed public inquiry came the next day.

The Premier and cabinet ministers went to Canberra. On 29 January the composition of the inquiry was announced. It would be chaired by a judge or a QC, with a marine biologist and a petroleum engineer as other members. Advisers to the two governments would put forward names to be considered. The inquiry was, said the Premier, to be 'as wide and independent as possible'; the main hearings would probably be in Brisbane, but the committee would probably visit the Reef.

This was not quite what we had in mind. Five members had been originally suggested. We had already urged that a marine biologist and a representative of the tourist industry should be included, at least, and we felt that the public, too, should be represented if possible by a conservationist. Much would depend now on who was chosen as the biologist member. As for the tourist industry, it had remained remarkably silent on the drilling question. The discovery John had made on the connection between Ansett interests and the oil industry put a possible interpretation on this. The State Minister for Tourism had never commented on the question of the possible effects of an oil industry on the Reef's tourist attractions. We had little hope that we would have much success in bringing in the industry on the side of Reef protection.

On 30 January we knew more of the result of the discussions in Canberra. The setting up of the inquiry would depend on the agreement of all oil companies to suspend drilling operations until the inquiry was completed. Twenty-six leases were involved. The Premier said he would ask the companies to which the state government had granted these leases, to continue to hold off from drilling; but he did not think that the state government could revoke the leases, if the companies insisted on their right to go ahead. He added that Queensland would insist that 'independent people' should be members of the inquiry – not only Commonwealth and state nominees. Government employees would be vetoed as members.

This meant that no officer of CSIRO would be a member. The Premier wanted the 'expert' members to come from overseas. The Prime Minister had won two points, however; the inquiry would be specifically into oil-drilling, instead of the wider terms the State was pressing for; and the state government would itself work for a suspension of drilling until the report was ready. A separate joint committee would be set up to investigate the Crown of Thorns problem.

We had publicly asked for certain principles to be laid down for the inquiry. We wanted it to hear evidence of all kinds concerning the Reef, including evidence from ecologists, biologists, meteorologists, oceanographers, economists and representatives of the tourist industry; and we wanted all evidence presented, and all the findings by the committee of inquiry, to be made public. We wanted the inquiry to give special attention, too, to the possible effects of cyclones on drilling rig operations and on operations at any established oil well.

The question of cyclones had been well illustrated, while I was in India. Cyclone Ada had devastated Ansett's own tourist islands, offshore from Proserpine. Wharves and shipping had suffered severely; steel structures such as windmills had been twisted and wrecked, and huge trees torn up and hurled far from their places. It had been a dramatic demonstration of the strength of the cyclones which so often hit Reef waters. When, like this one, they struck the coast, their violence was terribly impressive. The lesson had not been lost on the public, and many people had asked what would happen to offshore structures in another such cyclone.

As for the Ampol-Japex drilling, the Premier had given way on this also. He said he regarded it as a separate issue, not connected with the Barrier Reef argument, but would agree to the drilling being suspended, if the Ampol and Japex companies agreed on this, and provided there was no financial liability to the state.

The *Courier-Mail* commented that the inquiry promised to be Australia's trial of the century, with the verdict one of 'life or death' to the Reef. Also, the inquiry would unmask the state government's 'mystery men' – the experts the Premier was always quoting in his insistence that drilling was safe. 'Their names, their complete reports, and their questioning in the witness box will be awaited with interest.' Not only they, and the oil companies and other advocates of drilling, would have to present documented cases, but their opponents also.

What nobody had pointed out was that the only organisers for the opposition case looked like being ourselves, the three voluntary conservation societies. The Wildlife Preservation Society of Queensland, the Littoral Society, and the Save the Reef Committee had not enough money to pay for legal advice during the inquiry; if the Australian Conservation Foundation decided to join us, little would be added to our funds, for the Foundation itself was in financial trouble and making a drive for membership. We would have very little time even to write to the overseas witnesses, and those in Australia, whom we wanted to call, and none to pay their fares and other expenses. There would be even less time to organise an appeal for money for the purpose. We did not know when the inquiry was to begin, or how long it might take. Our own chief witnesses, Dr Grassle, Professor Connell and some others, were now overseas. Most of the work on the effects of marine oil spills on marine biology had been done in other countries, particularly the United States. If the inquiry was to be held in Brisbane only, we would need to bring at least some of our witnesses from those countries.

So far, Goliath had all the advantages, as well as all the money. The trial of the century would, it seemed, be a foregone conclusion.

And our other work, always exacting enough, was suddenly made even more complicated. There was a sudden, and we thought perhaps significant, rush of applications for sand-mining leases in the Cooloola area. The Wildlife Preservation Society, being the only incorporated conservation society in Queensland, was also the only one legally entitled to lodge objections to sand-mining and to appear in the Warden's Courts. The hearings were to be held in Gympie, many miles north of Brisbane. We would have to help out our local branches, which were taking much of the burden, and get our own witnesses to Gympie, as well as organising our case. On every front, we had far more than we could handle.

We decided to publicise the situation as strongly as possible. At the beginning of February, asked by the press how things were going on the conservation front, I blew my top. I called the proposed inquiry a 'gigantic piece of eyewash put on for the public's benefit', and stacked in favour of the oil companies. I went so far as to say it was a put-up job. The state government had refused any assistance to the conservation case; we had no resources and had to raise all our finances privately, through public donations. There was, I said, only one real answer to the Reef's problem – a complete moratorium on drilling and a thorough and lengthy biological survey. No public inquiry would meet the case. We had not asked for the inquiry, and without the biological survey it could serve very little purpose.

The Premier replied that the state and Federal governments had done all in their power to ensure the setting up of an independent committee. 'We anticipate the final committee will consist of a judge, a senior marine biologist, a petroleum engineer, and two other top men who really know what they're about. What more could conservationists ask for? It makes me wonder just what they're after.'

This publicity set off a wave of scepticism over the inquiry. Many people had not realised the problems the conservationist case was facing. But even more, they had not understood that, in effect, Australia had lost all power of decision over the Reef's future. It now belonged to the oil companies.

'Thus we have the absurd situation,' *The Australian's* leader said on 2 February, 'of the Prime Minister of this country and the Premier of Queensland jointly issuing a statement 'expressing the hope' that the companies concerned would suspend drilling pending the results of the inquiry. In turn, five of the companies so far have agreed to stay their hand but are emphasising that they see it as only temporary.'

The state Opposition chimed in. Given the terms of reference and the personnel of the inquiry so far known, said Mr Sherrington, the spokesman for the environment, he did not think the inquiry would serve any useful purpose. It was amazing that no scientists from CSIRO or the Academy of Science were included. The chairman would have no qualifications in science. If, as the Premier wanted, the inquiry was to start in a month or six weeks, it would coincide with the APEA Conference, when 'oil experts from all over the world will undoubtedly form a queue a mile long to give evidence in favour of drilling ... The vast resources of oil exploration companies will therefore be opposed to the limited resources of the conservation organisations.'

With all this doubt being publicly cast on the inquiry, it was certainly not fulfilling any hopes that the controversy over oil-drilling would die down. And now the accident-prone oil industry had yet another setback, this time involving human lives. In August 1969, there had been a huge gas blowout at an offshore drilling site in Joseph Bonaparte Gulf, in Western Australia. Ever since, the gas had been blowing wild and on fire while the oil-rig was being repaired in Singapore. Now in service again, the rig had been towed back to the site to begin plugging the well by drilling a relief site. This rig, the *Sedco Helen*, was now reported to have sunk, drowning nine men.

Every offshore disaster involving the oil industry now made big headlines. At first it was thought that the rig had gone too close to the huge gas escape, where the water, said shipping authorities, 'was no longer H_2O in composition' and where even a super-tanker would 'sink like an anvil'. This revelation seemed to contradict the geologists' assertions that gas was not a marine pollutant and that marine life would not be harmed by a gas escape. The point was taken by many; and writers to the press asked how far an oil blowout of the same proportions would have spread across the Reef.

In fact, the accident had not involved the gas escape, but a collision with a marker-buoy. But it drew attention once more to the frequency of blowouts, and letters to the press began again to demand that the Prime Minister take some real action to stop drilling on the Reef. From the oil industry's point of view, the accident could not have come at a worse time.

Meanwhile, Japex was still trying to get compensation for its costs before it would agree to halt its drilling. Since Ampol refused responsibility for this, the company approached the state and Commonwealth governments. Japex's Queensland manager said that the company was very sympathetic to the conservationists' views, and knew how important it was for Australia to preserve the Reef, but it was not fair or reasonable that Japex should bear the whole cost of deferring the drilling. (Certainly, as the press was now repeatedly pointing out, the Commonwealth could have refused to confirm the Queensland permits.)

And he pointed out that right up to the *Navigator's* sailing date, the Premier had been firm that the drilling should go ahead.

Was Japex indeed entitled to compensation, and if so, where did the responsibility for it lie? The state government handed Japex's letter to its lawyers.

And it now accused Mr Gorton of having opened the way for the companies holding drilling permits to seek millions of dollars in compensation. His call for a cancellation of drilling was, said the state government, a tactical blunder. 'We steered clear of this danger by imposing such severe conditions – while not prohibiting drilling – that the companies holding permits found it impossible to go ahead. The result has been that no holes have been drilled on the Reef since the Santa Barbara blowout.' Mr Gorton was responsible for the Japex claim, and the floodgates had been opened for others. If all the companies claimed compensation and succeeded, Queensland could be up for $30 million in compensation. (This was rather less than the 'hundreds of millions of dollars' reported earlier, by the way.) Thus, if any compensation was to be paid, it was the Commonwealth's responsibility, not that of the state.

The Mines Department official who made the statement added that it was 'a tragedy for Queensland' that the Japex drilling had not gone ahead. 'If Japex had found oil, the agreement stipulated that it had to be refined in Australia – and that clearly indicates Queensland. A new refinery would have had to be built in the state. We were on the way to breaking the hold the American oil companies have on the Australian market, and this would have led to a cut in the price of petrol to Australian motorists.'

Apparently the accusation by an unnamed spokesman, that the Queensland conservationists were tools of American interests in opposing a Japanese interest in Repulse Bay, was now explained.

The argument over compensation went on. A Labor Senator said that as far as anyone knew, Japex was not a party to any agreement with the Queensland government, and the matter of compensation lay between it and Ampol. Since Ampol had been the first company to move towards suspending the drilling, said another, this made neither the state nor the Commonwealth liable for compensation. The *Courier-Mail*, on the other hand, took the view that since the Commonwealth had approved the offshore permits, it should bear the cost of compensation. If Australians wanted the Reef conserved, they should be prepared for their national government to pay to have this done.

In the event, both governments were to refuse to pay the unfortunate Japex. But the argument going on above our heads did not much concern us. We had been too busy on our other fronts, and one notable victory had been scored.

At the beginning of February, the campaign had begun for a by-election in the state seat of Albert, on the Gold Coast. It had traditionally been a Country Party seat of the safest kind. Our own Gold Coast branch had been very active in the Reef question, and now it was out electioneering, with the Reef as its cause. We had not forgotten the Premier's challenge, before the state elections in the previous year: 'Is the Barrier Reef an election issue? I think not.' Now we had to prove that it was indeed.

But it would not be easy. The Labor Party was very willing to make the Reef an issue, but in that prosperous area it would be difficult to dislodge the Country Party on a question that did not directly affect the local electorate. The Labor candidate was a particularly good one – young, energetic and attractive; but he had not stood for Parliament before. The Liberal candidate, scenting which way the wind was blowing, had come out in favour of the deferral of drilling and of the public inquiry, as a personal view. The Country Party candidate, on the other hand, said that the Reef question was not an issue at all, and chose to fight on local issues and promises which would benefit the electorate.

If the Liberal candidate himself was wavering, we thought it indicated the uncertainty on the Liberal side of the coalition. One good push, and something might give.

The Premier himself was to open the Country Party's campaign. And the Party was holding a conference at Southport at the same time, to discuss Commonwealth-state relations. There would be an influx of about thirty parliamentarians and their supporters, including a number of federal ministers and the Deputy Prime Minister, Mr McEwen, himself a Country Party Member. This was indeed a concentration of politicians for a single by-election. We felt somebody was getting worried.

But even the political commentators were not expecting the Reef issue to be important. The Premier's prestige would be lessened if the election was lost, but nobody could imagine a Labor candidate winning that seat. The Liberal candidate was sure to win, if the Country Party lost; but the Labor preferences might decide which, if the election was close-run.

We had a large contingent to attend the Premier's meeting, not only from our own Gold Coast branch but from other organisations which had come out in support of the Save the Reef Committee. I went down on the day, and at a meeting before the campaign opening we decided our tactics. We had certain questions to ask: they would be directed at the Premier only, and we intended to see that they were answered. There would be no heckling during the speeches, but we would use the question-time thoroughly.

We scattered our forces through the hall, which was soon packed. It was clear from the beginning that most of the audience, even apart from our own contingent, had the Reef issue very much in mind. Above all, we wanted to know what the state government intended to do if the inquiry recommended against drilling.

There were few of the federal members of the Country Party at the meeting. We had half expected that the federal leader, Mr McEwen, might have been on the platform to support his Premier. There was a sense of isolation on the platform, however, with only the four state ministers.

In the event, some of the questions we had intended to ask were asked by others in the audience. The answers were predictable and predictably evasive. But the Premier made one serious error: he tried to share the responsibility for Queensland's stand on the Reef with the Liberal Party.

This angered the Liberals, whose hopes for winning the election were pinned on the personal stand by their candidate. Next day the Liberal leader, Mr Chalk, commented angrily. Not only was the guilt-by-association on the Reef bad publicity for his party, but on local issues, the Premier had taken the credit for anything done by the coalition government that benefited the electorate.

Our Gold Coast branch was hugely encouraged by the obvious feeling of the meeting. They kept up the pressure at later meetings, and reported to us that they were sure the Reef was in the minds of many people, whatever their political sympathies.

The Country Party conference at Southport later discussed the Barrier Reef question and widened it to offshore issues in other states, and to the question of tanker oil-spills. The Country Party had clearly begun to believe its image was being seriously damaged. There was a hurried announcement that the Queensland government would set up an interdepartmental committee on oil pollution. It would look into pollution from ships and tankers, as well as from oil blowouts.

Also, the government would establish marine parks. A 120-mile stretch of coastline from Innisfail to Cooktown might be a suitable area for the first of these. (No oil-drilling applications had been made here.

The question of whether oil-drilling would be allowed in such parks was not mentioned; but we knew, of course, that Queensland legislation could not override the Commonwealth's law, and that there were no provisions in this that prevented drilling.) The press reported that the Queensland government now felt that conservation was becoming a world-wide issue. The meeting at the Albert by-election had made it clear that the Barrier Reef oil-drilling could be a major electoral issue.

So we had won our point. Mr Bjelke-Petersen and his party had caved in, and the trades unions and the conservationists were now no longer an 'irresponsible minority'.

Whatever had happened behind the scenes in the councils of Cabinet and the parties, the Premier now seemed a changed man. At the second meeting in support of his candidate, he made a notable switch in his campaigning. His government, he said, was 'always the champion of conservation'. It had increased the area of Queensland's national parks to over two million acres. (The major addition it had made was the Simpson Desert area, in the far south-west of the state – an almost inaccessible region of sandhills, in which no mining interests were involved, and where few people could, or wanted to, travel in any case. This made up more than a million acres of the total of National Parks, and was more of a cosmetic addition to the percentage of land reserved as national parks, than a useful place for recreation.)

The government was planning for the first marine park in Australia. (This had still not been made, in 1976.) It had also appointed the first Conservation Minister in Australia. (This minister was, in fact, responsible only for the conservation of soil and water.) And the conservation of the great natural beauties of the Broadwater at Southport, in the Albert electorate, was 'very much in the forefront of our thinking'.

The by-election seemed likely to be very close. The Labor candidate had put a great deal of emphasis on the Reef issue and this was clearly influencing voters. The Liberal candidate had gone out of his way to express himself personally in favour of stopping the oil-drilling, and had openly attacked the Country Party's attitudes on the matter. This might just swing the balance in his favour. The Country Party candidate was driven to put local issues only as his policy and his answers on questions over the Barrier Reef were evasive.

On the primary count, Bill d'Arcy of the Labor Party was ahead – an unheard-of thing in that electorate. But he would need twenty-four per cent of the preferences for a win. The Labor vote had gone up from twenty per cent to over forty-three per cent since the election the year before, and the Country Party had lost thousands of votes. In the event, the Liberal candidate scraped home by a few preference votes. But for us it had been a notable victory.

To the press, the Premier refused to admit that the election had been a rebuff to the government or the Country Party. But even Country Party members were openly complaining that the party seemed to be flying in the face of public opinion on its conservation policies. The Premier's stocks were very low, and there was talk of a challenge to his leadership; but nobody was prepared to name a replacement for him.

Conservation was very big news, as a result of the by-election. The *Courier-Mail* ran a full-page feature article on Queensland's conservation problems, with interviews with conservationists. Len Webb was able to get in a public plea for his vanishing rainforests, and the 'confrontation' between conservationists, government and business interests was reviewed.

New members flocked into the Wildlife Preservation Society, and into a new body which had recently been formed, the Queensland Conservation Council. The increase in our own membership was so large that it strained our small organisational base in the office to breaking point. We had gained five new branches in a year. But, as we pointed out in the press interview, we were still a spare-time voluntary body without finances to handle the coming inquiry or any paid staff except a part-time typist to handle the work of our magazine. Everything we had done and were doing was at personal sacrifice.

The government was very much on the defensive, both on its record and on its future plans. The Mines Minister, interviewed by the press, said that he was himself a practising conservationist – he did not allow anyone to shoot wild duck on his property at Proserpine. The Government's priorities, he said, were the development of the country and the raising of its living standards. He called the criticism of the Reef drilling project 'unsupported'.

On 14 February, Japex announced that it would defer drilling in Repulse Bay. The contract on the *Navigator* had been terminated. The arrangements for the proposed inquiry could now go ahead.

World-wide, the oil industry was in the news, and none of it was good. Another tanker accident in Tampa Bay, Florida, spread a huge oil-slick over the bay's waters. And the coastline of Nova Scotia was fouled for forty miles by oil from a Liberian tanker that had run onto rocks and spilled over a million gallons of fuel oil.

132

Articles on the 'dying world' were still being commissioned by newspapers, and clearly the coming conference of the Australian Petroleum Exploration Association would not be held in an atmosphere of self-congratulation.

But already the press was showing signs of tiring of the issue. We could foresee that unless we could get real legislative action soon, the whole question of the future of the Reef might slip out of the news.

And, as an article in *The Australian* sadly concluded, 'pollution demands even greater attention than development; but while it should be rated higher on our list of national priorities, it is quite certain that this will not come to pass'.

10

From Torres Strait to Canberra

Early in February the Littoral Society and the Wildlife Preservation Society sent out a circular appealing for submissions and for information from marine research biologists and scientific bodies in Australia and overseas to help in organising and presenting the case for conservation of the Reef. After explaining the background of the inquiry, and its proposed terms of reference, we said: 'The financial resources of Australian conservation bodies are too severely limited to enable us to bring expert overseas ecologists to Australia. This means that, while the oil companies will be able to bring in an unlimited amount of evidence to support their case, conservation bodies will have to rely on voluntary submissions of written evidence from overseas biologists and organisations ... We would very much appreciate any information regarding the dangers of offshore oil-drilling and pollution that you can send us for inclusion with our submission. Alternatively, your organisation may prefer to prepare your own submission which we could present to the Committee for you. We must apologise for such short notice in this important matter, but unfortunately conservation bodies were not consulted when the tentative date for the hearing was arranged by our government.'

We also sent a second letter to Australian non-scientific organisations, asking them at least to express their views to the Committee of Inquiry. We hoped that the Committee would be taking note of public opinion and of the views of non-professional organisations like our own.

A third letter went to the members of our societies themselves. It, unlike the other letters, made an appeal for funds to bring at least one experienced biologist from overseas for the inquiry. 'A submission from such a scientist may make the difference between success and failure of our efforts in Great Barrier Reef conservation.'

We had thought of embarking on a widespread public appeal for funds, such as we would have had to make if we had gone ahead with the idea of issuing a writ. But there were two good reasons against this. We were more and more indignant over the inquiry and the government's failure to provide any money for the opposition case or for bringing witnesses. We did not feel it should be the job of the hard-pressed conservation movement to handle all this. The inquiry, after all, had been brought on by the Commonwealth, with unwilling state co-operation; it had been no part of our demands.

We still felt that for an interim holding measure the Commonwealth, could declare a moratorium and finance a really satisfactory long-term biological survey, and that in the final issue, it could take over responsibility for setting up a joint Great Barrier Reef Authority to look after the Reef. If we accepted the responsibility for getting all the witnesses this would be tantamount to accepting that the inquiry alone would be enough – and for us it certainly was not. We were not going to be driven into the kind of corner where, if the inquiry decided for want of expert evidence that it was all right to drill the Reef, we would have the blame for not organising a good enough case.

And, even more conclusively, there was our own lack of the kind of resources to handle an appeal on such a big scale, and to pay for the advertising it would need. We were already working day and night. The Wildlife Preservation Society had far more than it could manage; the Littoral Society was researching marine biological publications for information and for names of possible witnesses. We could not see ourselves handling a stream of small donations and letters from all over Australia, and perhaps the world, as well as all this.

We also wrote to the ACF asking them to consider importing a marine biologist. They had had an increase in their grant from the Federal Government in the previous year. But their membership was lagging badly. (Their total membership in January 1970 was only 2,284, while we, as an unsubsidised voluntary body, in one state had already topped 1,600 and our membership was going up fast.) The Barrier Reef symposium, with its over-cautious approach to a subject on which so many people were feeling so strongly, had blotted their copybook, and we were seen to be fighting the issue while they had done nothing, after Sir Garfield's letter to the Prime Minister, but await events. Even a symposium they had held at Adelaide in November, on the conservation of the Australian coast, had had no speaker on the Great Barrier Reef, and it had been very poorly attended. Their reluctance to get involved in public debate and to make public statements was probably hampering the progress of their membership drive, and we realised they would find it hard to commit funds to bringing out a scientific witness. They were undertaking a number of other projects, as well; and they were making no comments at all on the tumultuous events of the past couple of months. But they decided to consider what they could do.

John Büsst, in Bingil Bay, was planning to appeal to the Duke of Edinburgh and the World Wildlife Fund for finance to import scientist witnesses, and to make his letter public all over the world. This, he felt, would throw a ghastly limelight on both the Commonwealth and state governments in their whole relationship to the Reef question, and cause them at least some international embarrassment. We knew already of the concern of the International Union for the Conservation of Nature and of many overseas scientific bodies on the fate of this great world resource; there had been several major scientific expeditions to the Reef by overseas research bodies overseas, and all these might be persuaded to express concern for the Reef.

And there was the question of actual international legal authority over the Reef – still unsettled: no one knew where either the state's, or the Commonwealth's, authority ended. Any international row over the Reef would not be to the Commonwealth's interests.

The Duke of Edinburgh was President of the World Wildlife Fund; he was also coming to Australia shortly on an official visit. If things were not going better for us by that time we intended to raise so much protest during his visit that the publicity would be international anyway. He could scarcely refuse to give us a hearing on behalf of the Fund in a matter in which we knew it was already concerned.

John wrote that he thought he was 'really on to something' in this, 'Somebody should shoot me!'

He was hoping to go to Canberra before the inquiry committee was set up, to press for a conservationist to be among the members. Also, his Melbourne solicitors now had come up with a new legal line on Reef jurisdiction. Everyone was assuming that rights over the Barrier Reef beyond low-water mark were vested either in the Commonwealth, or in the state of Queensland, or possibly even in New South Wales. But the solicitors, after examining the situation, were not at all sure that they were vested in either.

There seemed to be no indication that the residual rights of the Crown, in the right of the Government of the United Kingdom, had ever been handed over in favour either of Australia or Queensland. If this could be found to be the case, the authorities granted for oil-drilling might be in any case illegal.

That would certainly give us a much longer delay than the inquiry possibly could. It had been speculated that the whole thing might be over in a couple of months; and the Reef's respite might end there, with the oil companies refusing to wait any longer, whatever the inquiry might recommend. In that case, nothing but a massive international trades union black ban would prevent the drilling; and we could hardly expect that overseas trades unions would take the same view of the importance of the Reef as our own were doing. But Dr Mather wrote to the press drawing attention to the lack of legislation protecting the Great Barrier Reef and the need for a protecting authority with 'the expertise, the resources and the responsibility to administer the area'. No government department existed to carry out protective measures that might be recommended by any inquiry. The Commonwealth and state should immediately pass legislation to ensure that conservation of the Reef became law, and to set up an appropriate authority.

There had been no legal problem in setting up the joint Commonwealth-state legislation under which most offshore oil-drilling leases were held. Joint legislation should be no more of a problem, she said.

This was an area in which the conservationists and the Great Barrier Reef Committee were really hand in hand for the first time. We still suffered, in their view, from being 'unprofessional stirrers', and generally awkward customers; the Ellison Reef case still rankled, we felt. But the first glimmerings of a melting came in Dr Mather's letter, with a graceful reference to 'energy and tenacity' on the part of 'citizen vigilantes'. We were more than willing for the Committee to join in, at least so far as to help us organise the case for the inquiry. Their contacts and their knowledge in the field of marine biology went far beyond ours; but we had been afraid that the influence of their geologist and public-servant members might still keep them at arms' length from the inquiry and its implications. Now we gladly took up Dr Mather's hint. In February, Dr Mather joined us in working on the list of witnesses whose names we should send to the Prime Minister as important in our case.

Also, we decided we would have to express our views on what the Committee of Inquiry should do. We wrote an open letter, sending copies to the Prime Minister, the Premier, and the press. In it we said:

We ask the Commonwealth and state governments to plan the forthcoming Committee of Inquiry on oil-drilling in the Great Barrier Reef waters so that it will have the widest possible scope for the consideration of fact, and not find itself confined to the consideration of argument between oil interests and conservationists.

To achieve this we believe it will be necessary:
* for the members of the Committee of Inquiry to obtain first-hand views of offshore oil industries in other countries, especially in California and Louisiana,
* for the Members of the Committee to visit areas where pollution is known to have occurred, such as Santa Barbara, in the company of recognised ecologists who are studying the effects of the spillage.
* for the Members of the Committee of Inquiry to obtain complete records of the four or more blowouts which have occurred in Australian waters, including the Bass Strait blowouts and those off Exmouth and the Joseph Bonaparte Gulf. We suggest that the Joseph Bonaparte Gulf blowout is particularly important to the subject of the inquiry, since it seems to demonstrate the difficulty of dealing with these problems in cyclone areas.
* We draw to the attention of both governments that a most difficult situation could arise in an 'open inquiry', where only the evidence that could be brought forward by interested bodies would be considered and where interested bodies were expected to have expensive legal representation.

Much important evidence could not be brought before the Committee by private bodies of citizens, though no doubt it would be made available to the Commonwealth Government on request.

With regard to expensive legal representation, this could result in pitting a handful of concerned bodies of private citizens against the resources and full-time energies of vast and wealthy organisations. Interested members of the public have already spent time and money lavishly to bring the question of drilling in Great Barrier Reef waters to the point where a public inquiry has been ordered. But there can be no justification for continuing this once the need for investigation of the situation has been admitted.

The protection of national property is the responsibility of governments, it is not a subject on which concerned citizens should have to appeal for funds from the public, in competition for charity with poverty relief and welfare organisations.

We are not anxious to 'protect' the Reef with our resources (limited against unlimited), our skill in negotiations (untrained against trained and highly paid) or our ability in marshalling overwhelmingly convincing arguments at short notice.

We urge that the Committee should have as its charter investigation of the facts of the matter, with the right to initiate research.

This was sent on behalf of the Save the Reef Campaign. Three weeks later the Prime Minister replied: 'I am in touch with the Premier of Queensland on matters in connection with the setting up of the Inquiry, and the suggestions you have made are being taken into account in the Commonwealth's consideration of these matters.' From the Premier we had no reply.

Meanwhile, we reckoned that if the worst came to the worst, we might raise enough money to bring Dr Grassle, and perhaps one other biologist, from the United States with donations we had been promised. But we did not know when the inquiry would start, or how long it would last, and we certainly could not pay their expenses here for long. We were already very short of time to get in touch with them and arrange matters, and we would have to know when any scientist we brought out would be heard. The inquiry began to look like a complete walkover by the oil companies and the state government.

At the beginning of March, the Commonwealth and the state were locking horns again. This time it was on the proposed Crown of Thorns inquiry. Mr Gorton had offered Commonwealth subsidy on a dollar-for-dollar basis for the inquiry. But the committee must be constituted soon and begin work; the Prime Minister told the Premier that he could only expect Commonwealth assistance with the problem if there was an inquiry. The Premier, on the other hand, did not want an inquiry at all. Instead, he wanted the state's own 'research programme' to continue. The state, he said, had already advanced $22,000 for this, and it would come up with a suitable programme for eradication faster than a long inquiry would. Cabinet was 'well aware of the dangers ... and anxious to begin fighting the problem as soon as possible'.

The membership of the other inquiry, into oil-drilling, was expected to be announced on 4 March, and to begin hearings four weeks later. A decision on the Crown of Thorns inquiry was now postponed until after this announcement.

On 3 March the state government announced that it was going to hold 'a wide investigation into environmental control', which would foreshadow the establishment of a big organisation under a director of environmental control. This would co-ordinate efforts in 'air pollution, water conservation, national parks development, fauna, flora and marine life protection, rehabilitation of mined areas, beach protection, and even landscaping of freeways'. The Premier claimed that this made Queensland a leader in conservation.

'A crisis of world-wide proportions is arising in this vital area of human responsibility,' he said. For a long time now the portents of this crisis have been apparent. We have seen them in the explosive growth of human population; in the single-minded integration of the powerful technology of the twentieth century with environmental requirements; in the deterioration of agricultural lands; in the insufficient planning of urban areas, and in the virtual extinction of many forms of plant and animal life. Cabinet decided that we should take a further major step forward – one which will place the Queensland government in the forefront of conservation planning among Australian legislatures.'

Lifting our heads for a moment from our own state-generated problems, we took time off to be stunned.

This extraordinary about-face, as will be seen, came to very little. But it was certainly timely. The very next day a 58,000 ton tanker, the *Oceanic Grandeur*, was reported to have struck a rock in Torres Strait and to be gushing oil over an area of six square miles.

This long-predicted accident had taken place like the climax of a succession of ever-graver events. The Ampol-chartered tanker was carrying crude oil from Indonesia to Ampol's Brisbane refinery. It had grounded in Commonwealth waters, just east of Wednesday Island on the western side of Cape York. The slick was spreading eastward and another tanker, the *Leslie J. Thompson*, was standing by to pump oil out of the holed *Oceanic Grandeur*. Five of the fifteen oil-tanks were pouring out oil.

The important pearling industry, worth two million dollars a year to the Torres Islanders, was threatened, and so were the local fisheries. The islanders largely lived by these fisheries and pearl farms, and the waters of Torres Strait supplied most of their needs.

What concerned us most was that it was reported that detergents were being flown to the spills, and that one of them was the same as that used in the *Torrey Canyon* disaster. This was Gamlen which had afterwards been shown to have been more damaging than the oil itself. The two detergents actually used were Gamlen and the vaunted 'non-toxic' Corexit. The report issued later said that 4,252 gallons of Gamlen had been used, and 1,760 gallons of Corexit.

The spills were dispersed with these detergents in an attempt to protect the pearl-culture farms nearby from being affected by surface oil. The pearl-shell fishery lay to the east and north of the spill area, and, as the Director of Harbours and Marine reported later, 'was not considered vulnerable to an oil slick if the slick was left alone'.

The Commonwealth Department of Shipping and Transport had been asked to locate stocks of detergent in New South Wales and Victoria, while the state Department of Harbours and Marine located supplies in Queensland and co-ordinated the arrangements for dealing with the spill. In the event, it was three days after the tanker had been holed before the detergent reached the area and was applied. There was also a long argument and delay before the second tanker began pumping oil from the damaged vessel. It was a delicate and difficult job, for the tanker was down by the head and might capsize or sink if her balance was changed. In fact, a second major spill was caused later in the pumping process.

In the event, the Torres Islanders were to suffer heavily from the disaster. Much of the oil finally moved out into the Coral Sea and dispersed there – with what final consequences will never be known – but much oil landed on some beaches. The main effects showed up much later. In 1970 there had been five large cultured-pearl farms in the area; it was reported in October 1972 that only two pearl-farms still existed in Torres Strait, and the detergent used was blamed for the loss of the others.

It was to be more than twenty-four days before the *Oceanic Grandeur* was finally cleared of salvageable oil from her damaged tanks, and during that time she was lying almost under the surface and was surrounded by oil from occasional spills. Miraculously, though two cyclones threatened during that time, the weather stayed fine.

The real tragedy was to the Islanders, who learned at first hand what a big oil spill and the use of detergents could mean to an island population which lived largely by the products of its surrounding seas.

On the same day that the news of the tanker accident made the headlines, Federal Parliament's March session opened. In the Governor-General's opening speech, there were two highly important announcements. The first was that the Government believed it was in Australia's national and international interests to have the legal position on the continental shelf resolved. The Government therefore planned to legislate to assume sovereign control over underwater resources of the shelf.

This was indeed a throwing down of the glove, on Mr Gorton's part.

One or all of the states was expected to challenge the legislation in the High Court, thus bringing on the case which Sir Percy had urged a year before.

The Premier said that he had known nothing of the proposed legislation. He was clearly angered by the announcement.

From our own point of view, the second announcement was just as important. The proposed oil-drilling inquiry had now been upgraded to the status of a Royal Commission.

This would make a great deal of difference to our problems. Royal Commissions, in Australia, can summon any person whose evidence they think necessary. Witnesses are examined on oath (as they need not be in inquiries) and they must produce any documents the Commission calls for. A Commission, then, could bring out facts that an inquiry could not, and witnesses would be immune from legal action on whatever evidence they might give – another point that had worried us. All in all, we drew much comfort from the inquiry's new status.

But there were still the problems of legal representation and of the expenses of our witnesses, and these seemed insoluble.

In the Queensland Parliament that day the Premier told the House that he still believed that drilling should proceed in Repulse Bay. Certain individuals, he said, had tried to make 'personal and political capital' out of what had been no more than 'an honest endeavour on the part of the government. It is to allay this degree of public concern that the state government has insisted to the Commonwealth that the opportunity be given for all the relevant facts to be presented to the inquiry.' Next day, he made a strong attack on the proposed Commonwealth legislation on offshore areas. He said the state had received no notice that the legislation would be introduced in the autumn session, nor that it would extend from the low-water mark. The legislation should not proceed until States Ministers had been able to meet the Commonwealth Minister for National Development.

But the tanker spill was the chief concern of the headlines, and the Premier flew up to inspect it. On his return, he said that Gamlen detergent was the most effective means available for dispersing oil spills. It might be harmful to marine life, but this was 'an acceptable disadvantage' in the vicinity of the tanker, where there were many miles of water having depths of fifty to sixty feet, and which was the 'logical site' to treat and disperse the spills before they could reach the established fisheries.

The newspapers now contained dramatic photographs of the lamed tanker and the great spills around her. Like the Santa Barbara photographs, they frightened many.

We had been circulating a second petition on the subject of the Reef. This time its wording was stronger. It asked the Queensland Parliament to withdraw immediately all permits for prospecting or test-drilling, to proclaim a complete moratorium on prospecting and test-drilling for at least ten years or until such time as a complete biological study of the whole marine environment associated with the Reef had been completed, and to agree to the setting up of a Commonwealth Commission wholly responsible for the administration, control and scientific study of the Great Barrier Reef and its associated offshore environment. This had been circulated in a great hurry, but the response was remarkable – almost everyone we had had time to approach had signed at once, and we had collected nearly 5,000 signatures in a very short time. The Premier, who had returned from the visit to the tanker spill, received it impassively, and we heard no more of it, then or afterwards.

Now began the main confrontation between the states and the Commonwealth over the proposed legislation. The Premier announced, on the same day that we presented the petition, that there would soon be a meeting of all States Mines Ministers to discuss action.

Significantly, and sadly, he was applauded from both sides of the House when he said that the state government was strongly opposed to the Bill. States feeling would always triumph over 'centralism', even over issues in which the states were seen to be acting against the national interest. It was a tribute to the strength of popular feeling on the Barrier Reef question that in this case, at least, 'states rights' took a back seat in the minds of most people.

On the following day, 11 March, the long-heralded and long-delayed *Navigator* slipped quietly into her Brisbane berth, to wait for a new assignment. The crew had a long holiday ahead. The rig stayed in Brisbane until it was re-chartered. Meanwhile, the trades unions kept watch on it to be sure that it made no unexpected moves.

Another major oil accident now jumped into the news. A month before, an offshore oil platform in the Gulf of Mexico had caught fire, and all attempts to put out the blaze had failed. Now it had been extinguished by dynamite blasts, but at the cost of an oil spill that was likely, said the report, to make history's largest oil slick. It was heading shorewards towards the oyster farms, and threatening a seafood industry said to be worth a hundred million dollars. If the well could not be capped, it might have to be re-ignited to get the spill under control again.

That day, too, the *Leslie J. Thompson* was reported to have loaded all the oil it could into its tanks from the damaged *Oceanic Grandeur*, and to be leaving for Brisbane. No other tanker was near to take over the job, and much oil was left.

On the same day, 12 March, the Queensland government rejected a report by an Academy of Science committee which had been independently investigating the Crown of Thorns problem. The committee included Dr McMichael, then Director of the NSW National Parks & Wildlife Service, Dr Frank Talbot of the Australian Museum, Dr Endean, and another marine biologist with Reef experience. The report was dismal. It said that the starfish plague would now be impossible to control except in a few areas selected for their tourist value. In the areas the committee had looked at, up to ninety-four per cent of living coral had been destroyed.

The report made no suggestions about the cause of the plague, but called for co-ordinated long-term research, and recommended an advisory committee to co-operate with the Academy of Science, the Great Barrier Reef Committee, the Australian Conservation Foundation, CSIRO, the Commonwealth Department of Education and Science, the University of Queensland, Townsville University College and the Queensland government. This would have given the scientific organisations and the Commonwealth a commanding lead over the Queensland government in recommending what was to be done. It did not surprise us that the report got a poor reception.

On 14 March, the State Justice Minister came back from the states conference, which had been attended by all the Mines Ministers and State Attorneys-General. He said he was determined to stop the Commonwealth steam-rolling the Bill on offshore control through the Parliament. Queensland, he said, was deeply concerned, because the state government had already issued about fifteen offshore authorities to prospect for minerals and had many others waiting in its files.

(It was these authorities that John had discovered in the Planet Metals prospectus two years before. Evidently they had been held up at Commonwealth level in the meantime.)

All the states governments were in a furore over the proposed minerals legislation. They condemned the fact that the Federal government had not consulted the states before announcing the legislation, as lack of courtesy. Demanding talks with the Commonwealth, they were asking for a co-operative scheme for territorial waters and the continental shelf, based on a 'pooling of legislative competence'. They had not yet discussed a High Court challenge.

Indeed, it was clearly not in the interests of the states to have the case brought on. Such judgements as already existed were not in their favour. Their only hope, then, was to bring so much pressure on the Commonwealth Government, already split between Gorton's few supporters and the Country and Liberal Party supporters of states rights, that the Bill would never reach the stage of becoming law and the High Court case could be evaded.

The fifteenth produced more news with a bearing on the Barrier Reef's problems. Another cyclone was reported to be developing in the Coral Sea, near Thursday Island, where the tanker lay alone and with at least 30,000 tons of oil still in its hull. The State of Florida was reported to be bringing a suit for $250 million against the Humble Oil Company over the Tampa Bay tanker spill in February, claiming that the oil damage to Florida's waters and beaches had damaged its tourist image. And the Louisiana Attorney-General was reported to be prepared to file suit against the United States Government and the Chevron Oil Company for the oil-well spill which was still pouring thousands of gallons of oil into the Gulf of Mexico, killing waterfowl and damaging oyster-beds.

We had more than enough bad publicity for oil companies and tankers to keep public interest in the Reef lively and articulate.

A lighter note came into the debate when Gerald Durrell, who was visiting Australia to work on a book, was interviewed on his impressions. He had spent a good deal of his four-month visit in Queensland and had been listening with interest to the Barrier Reef argument, especially to the Mines Minister's various statements. 'Mr Camm helped my trip greatly,' he said. 'He kept me in a state ranging from vastly amused to absolutely hysterical. The more pronouncements he made on the Barrier Reef, the funnier he became. He is my unqualified recommendation for man of the year – if it wasn't so damn tragic.'

He urged Australia to set up a federal department to co-ordinate conservation measures with action against pollution. Australia, he said, was well on the way to 'butching it all up' for future generations.

This gave us joy, something we had had little of in this affair.

Meanwhile, not only the state, but the Commonwealth Government also apparently refused to accept the Academy of Science report on the Crown of Thorns problem as final. And now the membership of the promised joint committee of inquiry into the Crown of Thorns was announced. Not one of its biologist members had been on the Academy committee. Instead, it was headed by a human geneticist from a New

South Wales university, and included Professor Maxwell, the geologist who had spoken harshly of conservationists. 'Another hand-picked one,' some people muttered. The committee's composition was hotly and publicly criticised by Dr Mather. It was not, she said, a committee of expert biologists, either on the Reef or on the Crown of Thorns. It had not the experience necessary to assess the information it would be given, and some committee members had 'preconceived ideas' on the Crown of Thorns. The GBR Committee had not been asked for recommendations or suggestions on its membership. Its appointment, she said, 'demonstrated markedly the alarming tendency of state and federal governments to ignore or avoid evidence from professional experts'. The committee would be a 'farce'.

The next event was to be the APEA conference. On 13 March, Dr Dale Straughan, an Australian biologist employed by oil companies to study the effects of the Santa Barbara leaks, had arrived in Australia to attend it. Interviewed by the press, she said she could not comment on the Reef with much authority, but she did not think it should be drilled. 'Spills will happen. If Australia could not possibly get its oil from anywhere else, it might throw a different light on it. I think the Santa Barbara area will recover ecologically, but we don't know the longterm effects.'

Useful as this was, we were sceptical about this hopeful start. Clearly the rest of the 600-plus delegates to the conference were not likely to support any such heretical statements. The oil industry was embattled everywhere, but only in Australia did it look as though an area it wanted to drill might actually be denied to it. It had drawn its ranks to present a united front. If Dr Straughan tried any funny business in her paper to be presented to the conference, we felt it would be remarkable indeed.

The APEA Convention, said *The Australian's* mining news reporter, was likely to be an occasion for 'considerable private recrimination'. Very little drilling was going on in Australia. There had been a long succession of dry holes in wildcat locations, and mineral exploration was now the main activity. The growing power of conservationists, as represented by the halt to Barrier Reef drilling, was only one of the industry's problems, but had made it unpopular. Moreover, petrol prices were to rise when Esso–BHP's Gippsland oil began reaching the market.

The addresses at the meeting gave us no surprises. The Premier, in opening it, said that Queensland's ambition was to become an independent oil state, and that many 'wild and erroneous' statements had been made on oil-drilling on the Great Barrier Reef. He did not say what these were, but surely some of those made by his own Mines Minister would have qualified.

APEA's chairman said that criticisms of the industry had been 'loud, derogatory and contradictory'. People in the petroleum industry were puzzled by the criticism. It was 'somewhat akin to such events as the next-door neighbour's dog walking all over one's prize petunias. Despite the annoyance, not too many of us would shoot the neighbour's dog, or even give it a kick where it would be felt most. If we are well-adjusted, we prize our relationship with our neighbours above such satisfactions. We know the neighbour did not intend his dog to do it, and we presume that he will try to stop his animal from doing it again.'

We meditated this interesting simile. Who was the neighbour? Presumably the Premier. Certainly we were the dog. The question of possible ill-adjusted reactions in the matter of shooting the dog was also interesting.

The Federal Minister for National Development, Mr Swartz, in what must have been the least popular of the addresses, told the conference that effective measures had to be taken to preserve the environmental balance. The problem on the Reef was the possibility of irreversible damage from a blowout. 'It cannot be denied that drilling for oil and gas in unknown areas is a very tricky business and that in spite of every precaution blowouts can occur,' he said. 'The Reef is a natural phenomenon of which we in Australia and people throughout the world have every reason to be proud. We must take all necessary action to preserve it.'

The vice-chairman of APEA tried to put up a defence. A flood of 'irresponsible statements' had come from people who knew nothing of offshore oil drilling, he said. Conservation had become a fashionable cause and the Great Barrier Reef had become a focal point. But the industry must defend itself against the 'thinly disguised foes of private enterprise' who had stepped into the opening created by the conservation cause.

Dr Straughan did not repeat her statement that she thought the Reef should not be drilled. Tourism, she said, did more damage to species of marine life than oil-drilling.

An American oil executive said that the Santa Barbara event was an extremely unfortunate accident which the media had reported in 'highly emotional terms'.

But the conference could not have been a happy one, and the bad publicity continued. On 18 March it was reported that Louisiana fishermen were filing a claim for $62 million against the Chevron Oil

Company for potential damages, and oyster fishermen in the Gulf of Mexico were putting in a similar claim for $8 million over the oil spill there. And on 19 March, the Federal Government rushed through emergency legislation to give it power to remove the rest of the crude oil from the hull of the *Oceanic Grandeur*. This legislation gave it power to bill the tanker's owners for the costs, and if these were not met or the tanker proved unseaworthy, to confiscate it.

Again, the threatened cyclone had not developed, but another might at any time, and the tanker still lay helpless, with no relief vessel in sight. Ampol was galvanised into sending the *Leslie J. Thompson* back from Brisbane immediately with salvage material and a repair team to patch the tanker up.

It was now fourteen days since the accident had happened. It was revealed that the *Oceanic Grandeur*, with a draught of thirty-eight to thirty-nine feet, had been travelling along a channel which was charted at thirty-seven feet – and that a depth of thirty-four feet had been found where she struck.

The state Opposition continued to hammer the point that the Premier had the largest shareholding in a company with a drilling concession in Barrier Reef waters. It was reported that he might be asked to give evidence to the pending Royal Commission, and that, if he appeared, he might be asked questions about his oil holdings. The Premier replied that he was unlikely to be called, and that his shareholdings would not concern the Commission.[6]

---

[6] In the event, the Premier gave no evidence to the Royal Commissions.

11

We Get Reinforcements

On 18 March the final composition of the Royal Commission was announced. It was now to consist of three members only: a judge as chairman, a marine biologist, and a petroleum engineer. No names were given.

We were not the only ones who had suggested that a conservationist should be included on the Commission, and we were not very happy with the choice of members. We asked for the membership to be increased from three to five, and suggested the names of a couple of overseas ecologists. It did seem rather remarkable, we wrote, that in all this long and protracted battle for the conservation of the Reef, the conservationists, who began it, had never once had an official representative at any inquiry.

Nothing came of this suggestion.

The ACF had written suggesting that the question of tankers plying in Reef waters be added to the terms of reference. We did the same.

An article in *The Australian*, on the day after the Commission was announced, examined what might be expected to come out of the Commission. Since the Queensland government still wanted the drilling to go on, and no doubt would feel the same way whatever the Commission's report might say, it seemed that little might come of it. There would be 'further threats from the unions and a justifiable outcry', and the whole issue would be back where it started. The article also criticised the terms of reference. These were:

Taking into account existing world technology in relation to drilling for petroleum and safety precautions relating thereto, what risk is there of an oil or gas leak in exploratory and production drilling for petroleum in the area of the Great Barrier Reef?

What would be the probable effects of such an oil or gas leak and of the subsequent remedial measures on the coral reefs themselves, the coastline, and the ecological and biological aspects of life in the area?

Are there localities within the area of the Great Barrier Reef and, if so, what are their geographical limits, wherein the effects of an oil or gas leak would cause so little detriment that drilling there for petroleum might be permitted?

If exploration or drilling for petroleum in any locality within the area of the Great Barrier Reef is permitted, are existing safety precautions already prescribed or otherwise laid down for that locality regarded as adequate and, if not, what conditions should be imposed before such exploration or drilling could take place?

What are the probable benefits accruing to the State of Queensland and other parts of the Commonwealth from exploration or drilling for petroleum in the area of the Great Barrier Reef and the extent of those benefits?

The Opposition had tried to include in these terms the probable cost that would be caused from oil and gas leaks. The Prime Minister said that if this were accepted, there would be a delay in starting the Commission's work, because the Queensland government would have to pass the same amendment. But he said that 'To all intents and purposes, this is covered by the terms of reference', and he believed the costs could be worked out outside the Commission itself.

The Australian article said that, on the terms of reference, there could be no other finding than that the danger of an oil or gas spill did exist and that the Reef could be threatened. If this was to be the criterion for stopping the drilling, then drilling must not happen. But only the Commonwealth could intervene to stop it.

As far as economic benefits from finding oil were concerned, on one economist's argument this could only happen if the price of petrol in Australia became cheaper with the discovery of local oil. In fact, the price would increase, as it had with oil discoveries elsewhere in Australia. Little employment would be made, because 'crude oil is a very small employer'; and even if there was some benefit, would it be worth it if there was the slightest risk to the Reef?

We, too, were feeling rather cynical about the Royal Commission, and not only for those reasons. It looked as though, short of a miracle, we were going to be up against the best legal brains the state government and the oil companies could hire, for what might be a long inquiry; and we had found no money for any legal representation at all.

Here the first part of that miracle began. A well-known firm of solicitors in Brisbane, Lippiatt & Co., came forward and offered its help to prepare our case, without fee.

For us, this was a very big step forward. The generosity of this firm, and the interest, sympathy and enthusiasm they showed when we met them, made a great difference to our own wearied forces. We showed them the scientific papers and other documents we had collected, and the list of witnesses we thought should be called, and after a few consultations with them, they thought we would have a good case to present. But a great deal depended on the Commission's decision on how many of them could be brought to Australia, at government expense, and on what the Commission's own programme of overseas visits of inspection and hearings might be. We might still have to find the expenses of some of our witnesses.

With the response to our own appeals to our members for funds, we now had perhaps about $3,000 in cash and promises for the purpose. This would cover some of our own immediate costs, and perhaps we could bring out one witness at our own expense. We still hoped for some help from the Australian Conservation Foundation, too.

But the Foundation was still keeping silence on the Reef issue, and its membership drive, launched in January, was going very badly. On the other hand, it now had a number of benefactor members at whom strong conservationists looked with some distrust, for they included firms and companies with mining and oil interests. As far as we knew, none of these firms had ever made donations to 'activist' conservation groups – they certainly had not done so to us.

There was also another matter to embarrass the Foundation in the matter of the Reef. Sir Garfield Barwick, their President, was also one of the High Court judges, and in fact had been one of the judges in the highly important La Macchia case. No doubt he was getting plenty of criticism for his dual role, from states-right supporters and others, and he could not very well be seen as taking an active role in an issue that involved states rights.

At the Foundation's Council meeting in April, we suggested that they should pay the costs of legal representation for the conservation case. They deferred a decision on this, but they had already decided that the Foundation should be represented before the Commission, and they agreed to discuss the possibility of paying the expenses of a witness from overseas.

The Queensland Trades and Labour Council, much encouraged by the popular approval of its stand over the Mackay drilling, passed a unanimous resolution declaring 'irrevocable opposition' to any form of oil-drilling or mining on the Reef, whatever the outcome of the Royal Commission might be. If drilling began after the Commission's report had been issued, said their president, the TLC would probably invoke a black ban against the drilling company.

This was great news. When the report was finally made, there was likely to be a rush by the oil companies, since none of them had publicly agreed to abide by the report's recommendations. Nor had the state government. The trades unions, however, held the key, and if they stood firm, the door would be closed on drilling, perhaps forever.

As for the Commonwealth Government, Mr Gorton, our only hope, was in political trouble, and might not be Prime Minister for long. If he fell, the next Prime Minister, whoever he might be, would certainly not go ahead with the offshore minerals legislation which was already embarrassing Mr Gorton himself. Unless an election brought in the Labor Party, we had little hope of Commonwealth action.

We ourselves had no let-up in our work. The case against the sand-mining leases in Cooloola was to be held on 11 May, and would go on for perhaps many days; we had our counsel's expenses to pay, and several of our executive members were to give evidence. All this meant more holes in our time and funds. But the Littoral Society and Dr Mather were working hard with our solicitors on the Barrier Reef case, and taking some of the strain off our own society.

We had decided to postpone the idea, put forward by John, of appealing to the World Wildlife Fund, through the Duke of Edinburgh, for funds to conduct our case. Until we knew the intentions of the Commission on bringing overseas witnesses, and on making visits to other countries, we had no idea what those expenses would be. Meanwhile, we felt, the Commonwealth Government had the true responsibility for the Reef, and Australia should foot the bill itself. The Duke had attended the ACF Council meeting in April, but neither his own address nor Sir Garfield's made special mention of the drilling issue. The Foundation was cautiously hoping that the Commonwealth's action in setting up the Commission was a move in the direction of a moratorium.

The Duke's own views on the Reef were well known. After a visit to the Reef, he had made a thoroughly curt reply to the Minister for Mines on the subject of the amount of 'noise' conservationists were making over the drilling issue – in fact, so curt that it had been hurriedly censored out of the Press. But no doubt, like Sir Garfield, he felt the situation was too delicate for him to be publicly involved.

Two questions troubled us: the terms of reference for the Commission, which seemed loaded against a negative answer to drilling, and the question of whether its hearings would be public. The Commonwealth-state joint inquiry into the Crown of Thorns was to be held *in camera* which had caused Dr Endean to protest. We wrote to the press demanding that the Commission's hearings should be open.

It is Australia, and not the present governments, which is responsible for the Reef's well-being. Therefore, all Australians have the right to know what dangers face it and what should be done to counter them. Conservationists, at least, have nothing to hide from any inquiry of this kind. They want the public to know the facts.

Mr Camm replied that the Crown of Thorns inquiry was a technical one, and he could see no disadvantage in not making it public. We waited to see if the Royal Commission, too, would be called a 'technical inquiry'.

The press agreed with us that the Commission's terms of reference were unsatisfactory. The lack of Australian experts on the Commission was criticised. When the membership was announced, it was to be headed by a former judge, Sir Gordon Wallace, with an English biologist, Dr J. E. Smith, and a Canadian consultant petroleum engineer, Mr V. J. Moroney, as the other members. Dr Smith had worked on the *Torrey Canyon* spill's effects on marine life, but this investigation had been sponsored by the oil companies.

Nevertheless, the Commission was a considerable advance on the original suggestion of a mere inquiry, and we felt optimistic that in spite of the narrowness of the terms of reference we could present our case, given the help we needed.

As for the oil companies, they were, said the chairman of APEA ominously, 'particularly pleased with the Commission's composition'. They would assist it in every possible way, and they were 'confident that 1970 would see the whole matter of petroleum exploration in waters near the Great Barrier Reef resolved in a manner satisfactory to all interested parties'.

Now we had another stroke of luck, and one that was to be crucially important. The Great Barrier Reef Committee decided to join the conservation bodies in appearing before the Commission, and to share the services of our generous legal firm. Since we now had a reputable scientific body with us, any scientists who might have been in doubt about appearing for unprofessional voluntary conservation bodies would be reassured. Approached by the Committee, they would be much more likely to agree to appearing, or to pooling their knowledge in criticising the oil companies' case. But the Committee, too, had no money for legal representation or to bring out witnesses.

Our solicitors were now representing five bodies – the three original Queensland conservation bodies, the GBR Committee and the Australian Conservation Foundation. The Foundation was considering whether it would brief a junior counsel to appear on its own behalf. If it did so, we decided, we would not share this representation. It was a question whether our own views and those of the Foundation would always be the same, in view of Sir Garfield's difficult position, and we wanted separate representation. Could we conduct our own case, with the background help and advice of Lippiatt & Co.? But we had no legal expertise, and no one with the time and independent income that would be needed to follow the whole of the hearings through. The solicitors advised against it.

There would be plenty of legal strength on the other side; the Australian Petroleum Exploration Association was probably going to employ a QC from Sydney, with a junior assistant. Planet Exploration Company and Occidental of Australia would be represented, so would the Queensland Mines Department. At least three strong legal teams looked heavy odds, however much help we got from our solicitors; they were indeed being tremendously helpful, but in a long series of hearings we could hardly expect that they could afford to spare the time and the money to carry right through.

But they had now approached the Queensland Bar on our behalf, and asked if any solution could be found. The Bar responded remarkably. Thirty-five of its members, including five QCs, were prepared to act for us without fee, for one week each during the hearings.

This was great news, especially on the publicity side. No such offer, as far as we knew, had ever been made by a Bar association's members on such a scale.

At the opening meeting of the Royal Commission, Mr Peter Connolly, QC, told the Commission that he, Mr F. G. Brennan, QC, Mr John Greenwood and Mr J. B. Thomas were appearing for the

Australian Conservation Foundation, and asked leave also to represent the four Queensland bodies which they understood were 'the principal conservationist bodies in the state'.

'The circumstances of our appearances are unusual,' said Mr Connolly, 'and we think it proper to present them to the Commission.' The four bodies had 'no funds for a protracted inquiry, involving inevitably a contest not only with the oil companies but also the government agencies'. Therefore the solicitors (Lippiatt & Co.) agreed to act without fee for these bodies throughout the proceedings. Under the circumstances 'the thirty-five members of the Bar had also agreed to act without fee'.

This was impressive enough news to make the headlines next day, as we had hoped. The more public sympathy and help were seen to be on our side the better.

The ALP's conservation committee was at work, too. A barrister appeared for the ALP also. They were launching a world-wide appeal for funds to pay for their representation.

The barrister for the Queensland Mines Minister at once objected. It was 'quite inappropriate' for a political party to appear before the Commission.

The chairman overruled the objection; it was an open inquiry, he said, and anyone who wished would be given a right to be heard. He had no doubt that anyone attempting to give 'irrelevant evidence', as the Mines Minister's representative had suggested, would be curtailed.

It was the list of topics which would probably have to be examined that looked daunting for us. They were outlined by the QC, Mr Woodward, who was appearing for the Crown Solicitor for the Commonwealth to assist the Commission. They were:

Topography (geographic limits of the commission); hydrology (depths of water, tide currents and wave action); meteorology (the influence of storms and cyclones and temperatures); basic ecology of the Reef area; geology and petroleum potential; historical (material of exploration in the past, and present leases); probable benefits (in the way of fuels, power, increase in employment opportunities and royalties); and existing legislation for oil exploration in Australia.

The 'heart' material to be discussed would be: drilling methods and technology; production and distribution (from source to refineries etc.); precautions which should be taken against oil and gas leaks; overseas experience and Australian experiences of actual oil and gas leaks, and the advances in technology in combating such incidents learned from past leaks.

Control of oil or gas leaks if they do occur; remedial measures to clean up spillage; and the possible effect of an escape of gas or oil on the coral life, coast-line and general biological and ecological life in the affected areas.

The Commission had adjourned until July, during which time it was to inspect some areas on the Barrier Reef. Soon afterwards, we had a conference with our solicitors and two of the barristers, Peter Connolly and John Greenwood. The list of background and 'heart' matters that had been given seemed to us to make it clear that this would certainly be a long job.

The free representation by a roster of barristers, they now told us, would be very difficult to organise; and it had a number of disadvantages. The number of subjects that would be examined, and the scope of the subjects themselves, meant that there could be literally hundreds of witnesses and submissions on a great many subjects. Hearings could go on for many months; some might be held in places other than Brisbane. The oil companies could afford to have their own barristers who, working on a full-time basis, would move round with the Commission from place to place and even overseas. Our own barristers could not possibly be expected to do this, and to keep up with the mass of evidence that would be piling up in the background, if they were not working on the case all the time.

We had begun to realise all this; but it seemed to throw us back where we had started. Our options were: to accept things as they stood, and present a case that was crippled from the start by the disadvantages of operating with a series of barristers who could not be expected to cross-examine with a full knowledge of the background of our case and the evidence already given; to appeal to the World Wildlife Fund – which would now take a long time to make a decision and get their help to us; to launch our own appeal, with all the work that that entailed as well as our work on this case and the Cooloola case as well (for the Wildlife Preservation Society at least); and to make our problems public in letters to the press, while we asked the Commonwealth for help in paying for full-time representation. With the advice of the lawyers, we decided to write to the Prime Minister and the Premier and to the press, and to ask the ACF to approach the governments too. (We already knew what the Premier's attitude was; and we had let the Prime Minister know of our problems in getting legal assistance, at the beginning of the year; nothing had come of that at the time.)

Our letter to the press read:

The generous offer by the Queensland Bar to provide the services of thirty-five barristers for a week each to represent the organisations preparing the case against oil-drilling in Great Barrier Reef waters for the Royal Commission into this question is very gratefully appreciated by the organisations concerned. But it must be made very clear to the public and to everyone concerned that this at best is only makeshift representation and that the case the organisations have to present will suffer severely if no full-time barrister is available.

This is so because (1) there is a tremendous amount of background information on the Reef area itself to be absorbed before any barrister could be conversant with the issues of the case, (2) no barrister entering the case for one week (in the midst of other important cases of his own) could be expected to make himself familiar with all this, plus the transcripts of evidence already given and cross-examination of previous witnesses, in the time allowed, (3) the Commission intends moving later in the case to Northern Queensland and perhaps to Sydney and elsewhere.

Our legal advisers are emphatic that only a full-time representation by a barrister or barristers could ensure that the case against oil-drilling is properly presented.

The organisations concerned and now preparing as best they can a case for the Commission are all private, spare-time organisations – poor but public-spirited. They cannot possibly pay for a barrister's services in this long and highly expensive Royal Commission. They are more than fully extended in finding and contacting possible witnesses, preparing evidence and researching information as best they can.

They represent a strong section of public opinion against oil-drilling in Reef waters. But their resources are inadequate to present this important Commission with the evidence and legal advice it needs to make a true assessment of the case.

We are appealing to the Commonwealth Government to provide this legal assistance. We appeal also to every reader of your newspaper who is concerned that the Royal Commission should make a true finding to help us in any way they can, and to press the Commonwealth Government to make legal full-time assistance available.

We are sure that the Royal Commission itself must realise that the situation is entirely unsatisfactory from the point of view of proper presentation of the case.

Surely it is ludicrous to expect that private unfinancial organisations such as our own should be able to meet the odds presented!

And, as an ACF Councillor, I wrote to the Foundation outlining what we were doing and suggesting that the ACF, as a national organisation, was the proper body to ask the Prime Minister for legal representation on our behalf.

I pointed out to them that since the ALP Conservation Committee was already launching its appeal overseas for money for legal costs, two such appeals would undercut each other. It seemed to us that the conservation societies, which represented a strong public interest in the Reef, ought to have some claim on public support through the Commonwealth. We ourselves were working for nothing – indeed, paying in our own time and money for having taken on the job at all. We had strong proof in the public opinion polls that people wanted us to do this; but it would not be fair to expect a few donors to take the place of the Australian people, or to divert donors' money from charitable appeals for the purpose.

I was about to leave for a six-weeks' lecture tour along the Queensland coast, which had been arranged long before and could not be put off. I would be back in Brisbane before the Royal Commission's first hearing on 14 July.

Meanwhile, the press had kept the issue well in the public eye. It was a measure of the way in which environmental questions were still holding the headlines. The *Sunday Mail* special commentator wrote, at the time of the opening of the Royal Commission and the Cooloola hearings, that the conservation cause in Queensland might give the impression to a casual visitor that it was flourishing. The impression, however, would be utterly false. If there was to be real action, a good deal of it would depend on the Premier's attitude, and he, and some of his Cabinet, seemed to want to 'lead this march from the rear'. He quoted Kenneth Galbraith: 'The tendency of the modern economy is increasingly to serve not the public convenience, but that of the more powerful producers. And it is convenient and in accord with producer interest to make automobiles that poison the air, and to dump industrial waste that poisons the waters, and to use chemicals that poison birds, fish and people along with the worms, and to allow the cities to engulf the countryside in an unregulated sprawl ... This is private enterprise. It can only be changed by public regulation of private enterprise and private land use. To all orators who evade this unpleasant truth, a delicately thumbed nose.'

That was a brave note which, in the press at least, was to die away in the mid-seventies, while the interests of the 'powerful producers' were to stand unchallenged except here and there. It was not to be the last time that 'environmentalists' were to get a fair hearing. But meanwhile, the world stumbled on its downhill road with little if anything actually being done to stand in the way of the 'powerful producers'.

On 24 April, *The Australian* had run an article surveying the different interests in the Reef battle, in its series *Is Australia Necessary?* The journalists who wrote it had talked with us in Brisbane, with John Büsst in North Queensland, with geology lecturers, the chairman of APEA, and others. What pleased me most about it was that for the first time, John Büsst's role in the battle for the Reef was given due public recognition.

Particularly now, with congratulatory letters pouring into the Wildlife Preservation Society's head office in Brisbane, I was embarrassed to be credited with too much. As the President of the Brisbane headquarters, I had generally had to make the public statements, write the letters to the press, do the television appearances and the radio and press interviews. For this, my 'curiosity value' was always useful – 'news' had to be made, and it was easier to make it with some figure well known beyond the environmental movement itself. We had to convince the Australian people that the Reef belonged to them and that it was in danger and we had to state our case for believing that it was indeed in danger. This was part of my job; but it was unfair to those who were working at least as hard behind the scenes and unrecognised. No doubt few people thought of the plain hard work and organisation by other people that went into the whole campaign.

But John above all had been the leader, the one to get in and work on the people who really mattered – the political leaders and backbenchers and senators, the unions and the political parties. He had organised help when we needed it; he had planned what steps we should take next, written letters all over the world, and in doing all this he had spent more than $15,000 of his own money, and neglected his own work even more than I had neglected mine. All the way from the Ellison Reef battle to the black ban, the halting of drilling and the setting up of the Commission, his influence had been at work. All the lively enthusiasm and resource he had brought to the battle, even when he was ill and expecting much worse, had kept us all inspired and at work, not only in the Wildlife Preservation Society but the Littoral Society too. Yet, apart from having been made an honorary life member of both societies, he had never had any recognition for it, though we had discussed this and had nominated him for one of the more important awards for natural history in Australia.

As Vincent Serventy said to *The Australian* journalists, 'John Büsst does not get all the publicity, but he master-minded the whole thing'.

Now at least, *The Australian* article gave him this credit, and quoted him in a 'star' position, in its interviews.

Otherwise the article did not contain much that was new, except two statements. One, by the Chairman of APEA, was that 'it is our job to convince the critics that conservation is led by our industry, not fought by our industry'. The other came from a geologist at Townsville University College, taking on with humorous enthusiasm the now predictable geologists' stand. The whole Reef, he said, should be mined and turned into concrete 'so that we can have some decent surfing beaches in North Queensland'.

I added that to my file of *bons-mots* on the Barrier Reef, along with some other good suggestions. One had been that the Crown of Thorns starfish should be trained to eat oil. The other was that the whole Reef should be drilled for oil, and the resulting petroleum put back in the form of plastic replicas of the coral that had been there in the first place. That would both create employment and form a new tourist attraction.

The statements by the Minister for Mines that had set Gerald Durrell giggling were there too, but they were unintentional, not intentional, humour. Perhaps the best of them was a suggestion he had made that, since oil was a protein, the fish should be able to eat it.

A week after our letters had been sent to the Premier and Prime Minister and had appeared in the press, a Liberal backbencher from Queensland asked the Prime Minister whether a scheme could be devised whereby the Commonwealth and state governments jointly could give financial assistance to the conservation bodies for effective and continuous legal representation. Mr Gorton answered that he could see considerable difficulty in endeavouring to present a case before a royal commission if the advocate were to change week by week. 'But I think it is of importance to Australia that all matters concerning this very significant inquiry should be fully argued and made known before the commission. Should there be organisations or a group of organisations which the commission feels could be assisted by the provision of an advocate before that commission then I would certainly consider approaching the Queensland government to see whether any action could be taken jointly along these lines.'

It was a month later that the Prime Minister sent a telegram to the conservation bodies. It read: 'In response to your request of 29th May I have today advised Lippiatt & Co. Solicitors Brisbane that the Commonwealth will meet their reasonable legal costs including counsel's fees for representing the interests of the five conservationist bodies before the Royal Commission on Barrier Reef Petroleum Drilling.'

The state had certainly not come to the party. But we had our representation, and we could go ahead with an easier mind. We had also saved money for the Australian Conservation Foundation, which was now sharing our representation.

I came back in time for the first hearing of the Royal Commission, a session which promised to be the first of a long, controversial and often highly technical and prosy series. I had not had time to see the Reef; but from the Büssts' veranda its calm lazy blue seas stretched far out to the Barrier, still unharmed, to all appearances, by the years. But I had seen and heard what was happening offshore. The Herbert River, pouring tons of tin-mining wastes into the waters from its estuary, had turned the waves along the beach at Cardwell a dirty grey. The rainforest clearing had silted creeks so deep that they had to be cleared with bulldozers to keep the cane-farm flats from flooding. The cane-mills and industries all along the coast were pouring out their wastes, the farms were sprayed with insecticides so heavily that the Barron River, I was told by a scientist there, was one of the worst polluted rivers in North Queensland after rain. Not a town but was putting undiluted and untreated sewage into creeks and rivers. Oil slicks from ships' tanks were often reported. The Reef was suffering; the illnesses of civilisation were already changing it. What chance had we of keeping it as it once had been, or even as it now was?

12

The Fight That Should Not Have Been Needed

The offshore minerals legislation which had so upset the states was still held up. The states were fighting strongly, and the Liberal Party was split. Mr Gorton's supporters in Cabinet were embattled. The former Minister for National Development, Mr Fairbairn, who had negotiated the joint offshore oil agreement with the states during Harold Holt's Government, was strongly opposing the Bill, and had a strong following. A number of Liberals were anxious to bring down the Prime Minister and replace him with a more amenable leader, who would not precipitate a head-on clash with the states on the issue.

Meanwhile, the international rush for offshore resources was coming ever closer, and the only Commonwealth legislation that Australia had was its fisheries legislation, the legislation for the control of living resources on the continental shelf, and the disastrous joint offshore petroleum legislation with its avoidance of any claim as to whose the authority actually was.

The minerals legislation was clearly urgently necessary for the international battles to come. But the states' opposition stood in the way, as did Mr Gorton's unpopularity in the Party. But his opponents were determined to bring him down, if it could be done without precipitating an election. The 1969 election had left them with a dangerously small majority, and any obvious lack of unity within the government would tell against it in any future election.

Mr Fairbairn argued that the government had 'betrayed' the states through their lack of consultation before announcing the proposed legislation. The Opposition took advantage of the situation to ask that all the correspondence with the states on the issue should be tabled. This would show whether the Commonwealth had in fact made any commitment to the states over mineral exploitation, when the petroleum agreements were being negotiated.

The Prime Minister said he had no objection to tabling all the documents on the proposed mining legislation, if the states had none. But the correspondence on the petroleum legislation had been between Mr Holt's government and the states, and had 'no relevance' to the question of minerals; and it was to remain in force anyway. (However, with the Barrier Reef issue so thorny, and with the Prime Minister's known attitude on the Reef, it seemed unlikely that the offshore agreement would not be changed, if the states lost the question of offshore ownership in the High Court.)

Under pressure from his own party, Mr Gorton made a temporary concession: he would delay the legislation until he had had discussions with the states early in June. But he made it clear that there would be no compromise. The meeting would only be to discuss questions of administration of the legislation, which he still intended to bring down as soon as possible. The Territorial Sea and Continental Shelf Bill, the first to be considered by the House, was already before it. The Offshore Mining Bill was to follow it; but this would be sent to the state governments for examination before the meeting.

The Prime Minister promised, too, that the states would have the lion's share of offshore mineral royalties. He offered to meet the states' legal costs in a High Court case. Some of them were reported to be considering bringing this. But since it would put at risk not only their offshore minerals rights, but others as well, it looked unlikely that any of them would co-operate and take the hook.

Indeed, it would have been hard to retain the offshore agreement on petroleum, once the question of authority had been determined, even if the Prime Minister had wanted to do this. It could have been held after such a judgement that the Commonwealth's agreement with the states was now void, since the states would have no legal existence in international law as far as offshore questions were concerned.

There was also the vexed question of legal jurisdiction over harbours and bays. At Federation, a bay had been regarded as an 'internal water' if its entrance was no more than six miles across. This meant that while Port Phillip and Sydney Harbour, both less than six miles from headland to headland, were certainly under states' jurisdiction, bays like Moreton Bay might not be so. All the port authority and harbour legislation and installations maintained by the states might be legally a Commonwealth responsibility in such bays. Repulse Bay, where Ampol's first drill was to have gone down, had been claimed as 'Queensland waters' by the Mines Minister under the twenty-four-mile rule; but it (and Princess Charlotte Bay, where Exoil-Transoil held a permit for drilling) might be held to be legally under Commonwealth control.

The question of what would be the jurisdiction on such bays and harbours under the new legislation was raised. The Commonwealth legal authorities replied that each case would be 'looked at on its merits'.

It was the high popularity of Mr Gorton's stand over the Barrier Reef that was most likely to give his opponents pause. If they prevented the bringing down of the legislation, and the Reef was drilled as a result, we reckoned that they might be fearing that the government would fall anyway. It was all very difficult.

For the Barrier Reef, then, three issues were important: the passing of the two Bills, the bringing of a High Court case to challenge them, and the question of jurisdiction over harbours and bays more than six miles across at the entrance. But a fourth issue looked like being even more important – would the Prime Minister be able to stay in office to see the legislation through, and who would succeed him, if he fell? Another election was likely to see the Labor Party in office, for the government was less and less popular on many issues.

The Opposition brought a censure motion against the Government on the question of the lack of consultation with the states over the minerals legislation. It was a narrow thing; some Liberals had been likely to cross the floor, and were only persuaded against it at the last moment.

The debate on the Territorial Sea and Continental Shelf Bill was postponed until after the states-Commonwealth consultation.

On 9 June the Queensland Minister for Mines and its Minister for Justice issued a statement saying that, no matter what laws the Commonwealth might pass, the state would 'go it alone' in the offshore sovereignty dispute, and continue to exercise jurisdiction offshore as it had done before. The position on the continental shelf beyond territorial waters, they admitted, was 'complex'; but the offshore petroleum legislation had 'solved this', and there was no reason why the minerals legislation should be any different. The division of royalties in the states' favour was not the point – it was the Commonwealth's attempt to extend its powers at the expense of the states that was worrying them. They said the Commonwealth, if it wanted to do this, should have conducted a referendum, as it had traditionally done, in any attempt of the kind.

There were two points to be considered here, we thought. First, the reluctance of Australians to change the Constitution in favour of the central government against the states was notorious. But it could well be held that, in issuing the offshore permits in the first place, the states themselves were acting illegally in terms of the Constitution, which stipulated that states could not alter their own boundaries. This was one ground on which we had considered issuing our writ, before the halt to the drilling.

It was now clearer than ever that the Royal Commission's report, whatever it might recommend, would in no way influence Queensland's determination to drill the Reef; nor would the decision of any High Court case. If so, we wondered whether Queensland would finally have to secede from Australia on the Reef's account.

Also on 9 June the Prime Minister told a Government parties' meeting that the offshore legislation would not be taken further during that session.

The Queensland Premier had been out of Australia on a six-week tour. He came back on 16 June to a new argument. Comalco had made a new share issue and had offered a number of shares to state and Commonwealth Ministers on a preferential basis. Six of the Queensland cabinet ministers had accepted allocations, outside the normal brokerage channels. Mr Gorton had told his own Ministers that they were not to take up the offer.

The Premier stepped off the plane to a waiting press barrage. He hotly defended the 'perfect right' of his ministers to hold shares in companies on which the cabinet had to make decisions. There was 'no conflict of interest'. He refused to reveal the extent of his own shareholdings to anyone. Responsible men, he said, would be 'driven out of government' if they had to divulge their shareholdings. *The Australian* commented that Mr Bjelke-Petersen had not shown such 'public anger' since he had defended his shareholdings in Exoil-NL earlier that year on a television programme.

'If Ministers are interested and prepared to back companies in this country, then these are the men we want,' the Premier said.

State Parliament was to resume on 21 July. The Labor Opposition was preparing to move a vote of no-confidence in the government over the issue of ministers holding preferential Comalco shares. The Minister for Works and Housing, who had been on a sub-committee which had inquired into questions of sand-mining in Cooloola, had revealed that he held 500 shares in a sand-mining company, though not one of those which had applied for leases in Cooloola. He had not declared this fact when he was appointed to the sub-committee.

The ALP State Secretary told the press that it 'had been revealed that more than half of state cabinet ministers held substantial shares in mining companies which dealt with the government', and that the ALP had 'confidential information' that a number of others also held such shares through nominee companies.

On 16 June the Queensland division of the Liberal Party held its convention again, this time in Townsville. The Federal Minister for National Development, Mr Swartz, was there to explain the Commonwealth's point of view on the new legislation on offshore minerals ownership. He said that a 'situation of misunderstanding and lack of knowledge' had been created. But the feeling of the convention was this time strongly pro-states' rights. Asked whether the Commonwealth decision to go on with the legislation was irrevocable, he did not give any clear answer.

The convention developed into a direct confrontation between the Queensland branch of the Liberal Party and the Federal Government. It demanded that there should be further discussions between states and the Federal Government until they reached a 'satisfactory agreement', and made it clear that this meant agreement 'within the terms of the party's total commitment to a federal system regarding offshore sovereignty'. They dismissed the argument that it was necessary, in international terms, to get the sovereignty question clarified before the international race for seabed resources meant that Australia was caught without any clear ruling on ownership. It seemed that the Barrier Reef was one thing – the strength of public opinion had infiltrated even the Liberal Party here – but that offshore mineral resources were another. The inconsistency of this did not appear to worry them.

That same day, the beaches of Townsville were heavily polluted by a big oil slick from an overseas ship discharging oil in the harbour. The lovely beaches of Magnetic Island were blackened and people swimming were covered in oil. The main slick missed the beaches because of the wind direction and went out to sea.

In Brisbane, at the same time, the Young Country Party conference ended with a unanimous vote of censure directed against ministers involved in accepting shares to which they would not receive entitlement as private citizens.

On 14 July, the Supreme Court in Brisbane was crowded with

conservationists, reporters and others interested in the first day of the hearings of the Royal Commissions. (Though they were usually referred to in the singular, the Commissions were, theoretically at least, a joint Commonwealth-State inquiry, and therefore should have been spoken of in the plural. The Queensland government had in fact budgeted $50,000 towards their cost, but later it attempted to evade paying this, saying that its provision of 'expert witnesses' from the Mines and other departments was all it was responsible for. It had therefore no share in paying the expenses of overseas witnesses, of the hearings and travel of the Commissioners, and of course none for paying the legal expenses on the conservation side. Since the Commonwealth was to bear all these costs, I shall go on referring to it, as most people did, as the Royal Commission.)

The old Victorian Supreme Court room where the Brisbane hearings were held was star-studded with QCs and junior counsel. The Minister for Mines, APEA, two oil companies and we ourselves were all employing QCs. The two technical assistant Commissioners sat each side of the Chairman, and we regarded them warily. Apart from the petroleum engineering consultant, whose likely sympathies were clear enough, the marine biologist, Dr Smith, had been employed by oil companies in investigating the *Torrey Canyon* oil disaster, which, he had told the Press already, had done less damage to marine life than was expected. He was Director of the Plymouth Marine Institute of Biology in England.

Mr Woodward, assisting the Commissions on behalf of the Commonwealth, said in his opening address that the Commission had been given a 'difficult and complex task'. It would involve studying a number of scientific disciplines: geology, petroleum engineering, meteorology, hydrology, marine biology and petroleum chemistry. This would be hard enough in itself, but they were being asked to consider a complex of some 2,500 reefs covering 110,000 square miles in this regard. Much of the expert information the Commission would need to hear was overseas, and witnesses would have to be brought to Australia; but how many should be brought could not be decided until the Commissioners had been able to find out how much evidence would be available in Australia. The limits of the Reef itself might have to be decided also.

He outlined a tentative programme for twenty-three witnesses to be heard in the first seven weeks' sitting ahead, but this would be 'less than half the total evidence'.

It looked like being a very long job, and a demanding one for us as well as for the Commissioners. Sitting on the hard benches and chairs

167

along the walls, we knew this would be the beginning of a new, but at least as hardworking, phase of the Reef battle as that we had just gone through. We had won the battle for public opinion and for a halt of drilling. What lay ahead was another matter.

Our own barrister, Peter Connolly, thanked the Commonwealth for its assistance in paying the legal expenses of the five conservation bodies. And we were launched on a job we considered would have been totally unnecessary in a sane society, and of whose outcome we were most profoundly sceptical.

In the event, the Royal Commission's hearings and investigations, including overseas visits, were not to end for over two years, and nearly a hundred witnesses were heard, some more than once.

Now we began a programme of weekly consultations with our barrister and counsel, and the solicitors' firm. At these, the evidence given was reviewed, we consulted on how to meet certain points, on what authorities and documents could be relevant; we made suggestions for possible new witnesses, and took away transcripts of the evidence for criticism. Most of this burden fell on Dr Mather of the Great Barrier Reef Committee, who faithfully circulated transcripts among scientists for criticism, comment and suggestions; and on our Vice-President of the Wildlife Preservation Society, John Roe, who as an engineering consultant could make valuable comments and suggestions on the technical evidence. The Littoral Society of Queensland's president, Des Connell, supplied references from his own library on petroleum chemistry and pollution and we all racked our brains and library references for whatever might be helpful. Freda McLennan of the Save the Reef Committee acted as secretary to the consulting committee, circulated summaries of evidence and got in touch with suggested witnesses. Since none of us were being paid for the job, and we all had our own work beyond the Royal Commission, we had limited time, but we had such invaluable friends and guides in our legal team that even when we were unable to give much help, we were wholly confident that the case was in the best of hands.

Indeed, we could not have been better represented. The amount of extra work that they and their staffs did for the case, over and above the immediate job, meant that the 'reasonable legal costs' the Commonwealth had agreed to pay were not going to come near paying for the work they did on the Commonwealth's behalf and our own. If those costs had ever been reckoned out, they would have been sky-high – but they were not.

John Büsst, ill and overstrained, came down for some of these early consultations. But it was not possible for him to stay in Brisbane, and, back at Bingil Bay, he fretted in the absence of news. We had no time to keep him informed; and the transcripts for the conservation side were few and constantly being used, so that he had to rely on the press reports for much of the information. I wrote him whenever I could and whenever his special knowledge or advice was needed; for the rest, the press reports began to become sketchier, scrappier, and more and more bewildered by the sheer mass and technicality of the evidence as it piled up. From eager attendants at the hearings, the press finally drifted away; and some of their reports were necessarily ill-informed on the background of the case as it went on.

Indeed, hearings that went on all day, week after week, and cross-examinations that largely depended on evidence given by previous witnesses, or statements and documents that were long and technical, meant that the press, however willing, simply could not afford the time or the journalistic expertise to follow the hearings through. So gradually the Royal Commission dropped out of the news and the hard behind-the-scenes slog dragged on largely unnoticed. Occasionally there was a complaint at the sheer length of time that the hearings were taking; occasionally there was a statement from the Mines Minister about the disadvantages Queensland was suffering from the lack of an offshore oil industry and the general slowing down of oil exploration; occasionally the unions reaffirmed their determination to apply a black ban if any oil company jumped the gun. Otherwise, the emphasis in the press shifted from the oil-drilling problem, to the Crown of Thorns.

It would be impossible, of course, to give even the briefest meaningful outline of the Royal Commission's hearings and evidence. We learned much from the witnesses, but none of it materially changed the original positions of the oil companies versus the conservation interests. The oil companies, the geologists, and the Mines Minister still maintained that, though oil spills and leaks and pipeline breaks might not be avoidable (and indeed the overseas evidence certainly corroborated this), the Reef ought to be drilled; the biologists, ecologists and conservationists still maintained that drilling ought not to be undertaken in the shameful state of ignorance about its possible effects on the Reef, and that the Reef's value was not as an oil reservoir but as a living whole. Around those unshakable positions revolved the witnesses on the technical and economic aspects of the inquiry, on oil pollution's effects on coastlines and marine organisms overseas, on the meteorology of Reef waters, on the great tankers still manoeuvring their way through the Reef's narrow and dangerous channels and weather. Much that came out in the evidence, and in the cross-examinations, made us feel that whatever the outcome, the setting up of the Royal Commission had been more than justified. For instance, it became very clear that the Ampol-Japex operation, so vaunted for its safety precautions, was in fact very lacking in these, to the extent where the drilling muds would have been dumped overside in plastic bags. The proposed blowout preventer test pressures were so minimal that even a member of the Mines Department staff said that he had been surprised by them. And one highly important blowout prevention measure – the hydril – was apparently not to have been used at all.

It was permissible to wonder why the *Navigator* had taken so much time being fitted with the latest prevention measures. It was also, it appeared, permissible to wonder about all those assurances from the Mines Department and the Premier.

Queensland, it emerged, had no body of offshore drilling regulations in operation at all, though a draft had been prepared. Offshore drilling would therefore have had to be conducted under the ordinary regulations for drilling on land. The Queensland government apparently preferred to negotiate with the oil companies directly, rather than impose a code of regulations. Even the draft regulations, when compared with the United States' regulations, were highly inadequate – and we knew what had happened offshore of the United States.

Other questions came out. The Repulse Bay drilling was to have been conducted from a ship, not from a platform, in a region of very high tidal variation and cyclones where a fixed platform would clearly have been safer. All these factors, the Mines Department official had to agree, were not exactly satisfactory.

If we had done nothing else, we had prevented Mackay, and the Repulse Bay area, from being subjected to that particular set of risks.

This, and other evidence, was a more or less forlorn comfort in that if – and, as we feared, when – the Reef finally went down to the oil companies, better procedures might be installed and stronger conditions imposed than would have been the case if drilling had not been halted and the Commission set up.

There were comic moments too – as when a Mines Department geologist claimed that the Reef might be said to be suffering from a 'plague of corals'.

But basically the trouble with the conservationist case was exactly the lack of research into the ecology of the Reef that we had pointed to as justifying a major research programme. There was contradictory evidence on whether coral would be harmed by exposure to oil; there were demands from the Chairman that experiments be conducted on the Reef itself; demands that were strongly contested by the Great Barrier Reef Committee's scientists and ourselves. A proposed experiment, which would have been ad hoc, short term, and would have been vulnerable to scientific criticism, was abandoned; the state government's marine biologists then conducted one of their own, tipping eighty gallons of oil on Wistari Reef, to the strong condemnation of scientists.

Indeed, the question of harm to corals was not our main point; rather, we emphasised the danger of long-term ecological change to a rich and varied marine fauna and flora whose inter-relationships were unknown to science. But only marine ecologists with experience of such change in the Barrier Reef environment caused by oil pollution could prove the likelihood of such change beyond all doubt – and they were precisely the witnesses we could not find, because they did not exist.

What did emerge for certain, even from some overseas witnesses who might have been expected to be apologists for the oil industry, was that the offshore oil industry, once established, could do more lasting damage to marine life through small but continuous spills, detergent treatments, discharge of the water and muds used in drilling operations, and other kinds of pollution, than even single large spectacular oil accidents would do. The mazes of pipelines, of shore and offshore installations, of dredging and filling operations for these with their consequent silting, the channels dredged and blasted and the changes they made in tidal and current patterns, the accumulation of toxic chemicals on bottoms and shores (as was happening in places like Louisiana) – all these seemed inseparable from offshore oil industries. The loss of birdlife, an essential part of marine ecosystems, was beyond dispute in oil spills; those could be seen and counted, and seabirds could be observed. But the bottom-living organisms that are the basis of most marine life, and the plankton that replenishes it, could be suffering even more; and little work had been done on this, or on its final effects. Nor could it be easily observed, as could the more spectacular loss of birds in oil spills.

It was this – the ugly sad succession of little 'unimportant' unspectacular happenings, the small breakdowns and leaks and spills, the pollution from installations ugly enough in themselves, that we feared, and fear, most for the Reef. Those brilliant, white sand beaches and cays, those clear waters, those coral gardens with their millions of inhabitants – what would happen to them when the increasing forest of oil rigs, pipelines, shipping, spread over the Reef? Gradually, year after year, the blight of change and death had altered and lessened the beauty of the Florida reefs and the rich variety of the life of the Gulf of Mexico. That, too, was already planned for the Great Barrier Reef, which biologists, in awe at its wealth of living things, called simply '*the* Reef'.

It was this that needed to be pictured and told. But as the time wore on and the words piled up, the fate of the Reef lost its immediacy for people. Other problems concerned them now. Then came the long delay in bringing out the Commission's Report.

It will be useful to summarise what had happened meanwhile at government levels. The new Labor Government which came in at the end of 1972 had strong views on controlling overseas ownership of minerals and oil, and the offshore legislation was crucial to this. The new Minister for Minerals and Energy soon made it clear that an entirely new game was now in progress. In December 1973, the Government managed to enact part, at least, of the new Seas and Submerged Lands Bill, which was substantially the same as the legislation Mr Gorton had proposed. (He crossed the floor of the House several times in the voting.)

But the Senate removed the mining code in the Bill, under which the Commonwealth would have had rights of administration, thus leaving the states with their previous rights in this field. The Act now simply asserted that sovereignty over the sea and seabed resources below low-water mark belonged to the Commonwealth.

The Liberals had allowed this in an about-face on their previous stand, largely because the prospect of the vitally important Caracas conference on the seabed and its resources to be held in June 1974 had convinced them that it was necessary to establish ownership, and it did not appear that states' ownership could possibly be recognised internationally. But states' rights, within the closer limits of administration, need not be challenged.

Nevertheless, the states reacted strongly. They referred the question to the Privy Council. They failed in this bid when the Queen declined to accept the submission. Nothing was left but the appeal to the High Court, which they had so long feared.

Then in 1974 the states of Queensland, Western Australia and New South Wales issued writs to restrain the Commonwealth legislation. The Queensland writ was much more fundamental than the others. It requested the Court to declare that the Commonwealth had no power to make laws affecting the continental margin off the Queensland coastline. This was largely based on an argument being put forward in the United States, where several seaboard states refused to accept the jurisdiction of the Federal authority from the three-mile limit to the edge of the continental shelf. Maine and Virginia were claiming jurisdiction on 'historic rights' as far as 200 miles from their shoreline – and it was ironic that they were doing this not just, like the Australian states, to get rights over minerals and oil, but also to prevent the environmental problems of pollution and oil spills from industry that Louisiana was already suffering.

Much was at stake, including an international agreement with Indonesia over the border of the territorial seas – which related to oil rights on the North-west Shelf – and several international conventions into which Australia had entered or might enter. There was also the crucial question of the Torres Strait, where the border between Australia or alternatively Queensland – and the future nation of Papua New Guinea – was being hotly disputed. The Torres Strait and the Gulf of Papua were regarded as potentially oil-bearing. The shallow waters and their coral reefs were possibly to be regarded as continuous with the Great Barrier Reef itself. The Torres islanders, still suffering from the results of the *Oceanic Grandeur* spill, were strongly against oil-drilling there; at the same time they did not wish to be transferred from Australian sovereignty to that of the new and yet untried Papua New Guinea administration.

All these matters, and others, would hinge on the results of the High Court case. The late appearance of the Royal Commission's report, however, was to mean much. The High Court case would not be brought on until March 1975, by which time the Report could have been thoroughly studied. The question of what actually constituted 'Great Barrier Reef waters' was still unsettled. They could begin at low-water mark and stretch to the edge of the outer Barrier, if the High Court upheld the low-water boundary of the states. Their northern limit was also still debatable.

All in all, the High Court case would hold the fate and future of the Reef; and until it was decided, the Commonwealth could do little in the matter of altering the offshore petroleum agreement.

By that time, Thor Heyerdahl's second cross-Atlantic expedition in his reed ship had revealed that all the way across the Atlantic great floating lumps and chunks of solidified or still liquid oil and other wastes signalled what was happening to the world's oceans. More oil slicks had been spilled on the Reef – far more, probably, than ever were reported; and not only on the Reef, but all over the Pacific, tankers and ships leaked, discharged their tanks, and added more damage to the pollution pouring from human settlement and industry. The figures for the amounts of oil spilled, or discharged, or simply released from the sewer-pipes of cities, were given to the Commission; they were frightening. Oil, that 'blessing to be cherished', that source of most of the energy that drove civilisation and industry, was being returned to the seas which had been its origin in toxic and increasing quantities everywhere. Oil would not last for ever, but by the time the last oilfields would be found and exploited, the living seas and coastlines might themselves be dead or dying.

But oil was king; its users and industries ran the world. To stand in their way, there was nothing but public opinion on the Reef; that, and the kind of publicity we had been able to muster to keep it alive and at work. Would it be possible, when the Report came out, to muster that force again as strongly as we had been able to do before the Commission was set up?

During those four years, many things happened to make us fear that this might not be possible. The leap in oil prices imposed by the Middle East oil-producing companies made every drop of oil seem precious to most people, whatever sacrifices might be made for it.

Whatever its goodwill, the new Whitlam government could not introduce and implement legislation to declare the Great Barrier Reef a marine national park until the Report of the Royal Commission had emerged and been studied, and until the challenge by the states on offshore ownership had been decided in favour of the Federal government. Meanwhile, economic problems were taking over from those of conservation as the main issues in the public mind. The Whitlam government was dropping away in popularity as inflation increased. Would it even be allowed to govern for its full term?

When the Summary and Recommendations of the Royal Commission finally appeared, at the end of November 1974, they set off very much the train of consequences that we had feared. The Commission was, predictably, split in the view it took, the Chairman maintaining that 'all drilling throughout the Great Barrier Reef area, including the area east of and adjacent to the outer barrier, should be postponed and be planned and permitted only after the results are known of the short and the long-term research recommended in the Report.' The two other members of the Commission themselves took differing views of allowing drilling in various areas, but felt that it could be permitted in certain parts of the Reef's waters under certain stringent conditions.

This is not the place to comment on the Report itself; this would be a huge task, involving the examination of much of the documentary evidence on which the Commissioners based the varying views they expressed. Enough to say that it placed much less emphasis on the biological evidence, and the economic evidence, we had presented, than we had hoped, and indeed, even though it laid stress on the low level of biological and ecological knowledge of the Reef and its organisms, the contradictory nature of much of the evidence given on the effect of oil on marine ecosystems and the need for more intensive and extensive research, it seemed to us that the Commissioners had too much accepted the view that their terms of reference were to state where and how the Reef could be drilled – not whether it should be protected from drilling.

In any case, the appearance of the Summary and Recommendations set off the train of consequences we had feared. First, the Queensland government staked an immediate claim within the three-mile limit offshore from the coast and around the islands and cays of the Reef. This was done by declaring all these waters a 'marine park'.

This apparently innocuous move actually did nothing to protect any part of the Reef, its waters, or the coastline and seas within the three-mile limit. The classification of 'marine park' is in fact conservationally meaningless in Queensland law. It merely defines an area which the government considers is within its jurisdiction, and within which it may, if it chooses, later declare a 'marine national park'. Even this latter classification does not protect the area from oil-drilling, nor from commercial and line-fishing. Only sedentary marine organisms are protected in such a 'national park'; and the state government's record of policing and managing even its regulations against shell-collecting is not encouraging.

A committee to advise on creating such 'marine national parks' within these 'marine park' areas was set up. This consisted of state public servants and no provision appeared to be made for independent scientific advice to the Committee.

In any case, little further biological research had been financed or carried out on the Reef since the setting up of the Royal Commission, except for some work on the Crown of Thorns problem. The demand, voiced by the Great Barrier Reef Committee's scientists and the Academy of Science report, for such a programme to be carried out before any more development was started, had gone almost unanswered. The first stage of the Townsville Institute of Marine Science, announced by Mr Gorton as Prime Minister, had gone no further than the employment of a few researchers for its staff; the buildings were non-existent and the original site chosen had been vitiated by the pollution caused by a large nickel smelter in Halifax Bay. The site was now to be moved to Cape Cleveland. Clearly, the biological knowledge of the Reef that was needed was simply not available.

When the Report was released, Queensland was again facing a state election. Most people had been lulled into believing that Reef drilling was no longer an issue. Conservationists organised a meeting to discover the attitudes of the various parties on conservation; the coalition parties were represented only by one member of the National (formerly Country) Party; no Liberal accepted the invitation. It was significant, perhaps, that in 1971 the Liberal Party's Queensland conservation committee had declared itself in favour of the Great Barrier Reef's becoming a national park – not a series of small national parks – and that this was the declared policy and intention of the Whitlam Government, whose popularity in Queensland was rapidly waning.

The National Party's spokesman said he had no doubt drilling would not take place. 'The Commonwealth doesn't want it, the Premier doesn't want it, the trades unions don't want it,' he said. He pointed to the Premier's recent announcement as a proof of this. When he was told that neither marine parks nor marine national parks were protected from drilling under Queensland legislation, he replied simply that we 'could trust the Government'.

Campaigning in Queensland for the State Labor Party, the Prime Minister promised that the Australian government would create a marine park 'based on the Barrier Reef' in which oil-drilling would not be permitted. Its boundaries were not specified, though it was promised that the park would be very large. The legislation was then receiving its 'final touches' he said, after the Royal Commission Report had been studied.

Mr Connor, as Minister for Minerals and Energy, had already announced that the Barrier Reef area would be specifically exempted in applications for oil-drilling leases. This statement was made in March 1974, in an address to the Australian Mining Industry Council.

The long-postponed challenge to the Seas and Submerged Lands Act, which claimed offshore rights for the Commonwealth, was heard by the High Court during 1975, the case beginning in March. But it was clear that the states would not easily accept any verdict that they did not like. Indeed, by the end of 1974, states were already breaching the joint offshore agreement on oil exploration by renewing offshore drilling permits without consulting the Australian Government. Western

Australia, Victoria, and South Australia had issued permits, on their own authority, for five more years' exploration. In Queensland, the situation was complicated, however, by the trades unions' continuing ban on any drilling activities in Great Barrier Reef waters, by the sensitivity of the public on the oil-drilling question, and by Mr Connor's announcement. But the legal basis for the latter would not be clear until the High Court judgement, and the offshore agreement was still applicable.

On 11 December 1974, with the Queensland government re-elected by a huge majority – largely in reaction to the unpopularity of the Whitlam Government – and the National Party well in the ascendant in the coalition, the first major conservation bill of the Australian Government, which had been passed in the House of Representatives, came before the Senate. The Opposition Parties in the Senate amended the Bill so as to prevent the Federal Government from assuming control over national parks without the permission of the State concerned. This meant that the proposed Barrier Reef National Park could only be made after the High Court judgement, assuming that this went in favour of Commonwealth offshore ownership.

On 15 December, it was reported that the five oil exploration companies whose operations had been closed down before the opening of the Royal Commission hearings were resuming oil search in the Gulf of Papua. These waters were now under the control of the new Niugini Government.

Meanwhile, governments were studying the large two-volume Report of the Royal Commission. The split in the views of the two overseas technical members of the Commission and the Chairman's own dissenting view made the full report a problem to the authorities. Dr Frank Talbot wrote, in a letter to the press:

> The Australian Government's National Parks Bill, and an order in council for the Queensland Cabinet relating to marine parks, are both currently being considered and both will have a major effect on the future of the Great Barrier Reef.
>
> Queensland has indicated it will declare parts of the Great Barrier Reef as marine national parks, and the Australian Government has committed itself to declaring the whole reef a national park – at least that part of the reef to which it lays claim, which will include all the area below low-tide mark, if the Seas and Submerged Lands Act is upheld by the High Court early next year.
>
> Nor are the two governments co-operating, as the Queensland

Government apparently does not reply to letters from the Australian Government, and currently the political posturing has reached personal name-calling between the Federal and Queensland leaders.

Although it is encouraging to see governments fighting for, and not against, declaring national parks, particularly neglected marine parks, there are nevertheless inherent biological dangers in the actions proposed.

Biologists are now realising that the reefs vary fundamentally from north to south along the Great Barrier Reef, that many species become regularly lost to reefs by extinction of local populations, and that the richness of the reef is maintained by constant recruitment of larvae and juveniles, predominantly from north to south.

In addition, the islands and reefs are closely related, with herons, terns and other birds feeding on reef foods, and turtles breeding on the cays. The only way to protect and manage the reef, as it inevitably comes more under siege from increased land use, and from open ocean pollution, is to consider it one biological whole.

Piecemeal declaration of parks, or separation of islands from reefs, makes biological nonsense and is bound to have adverse long-term results ...

Australia holds this precious heritage in trust for all mankind; we must surely expect political and administrative sanity between the two governments for its management and control.

This is certainly not obvious at present.

As a financial newspaper pointed out, control of the areas recommended for drilling by the Royal Commission Report was a main area for conflict between Commonwealth and state. Exploration companies had held their rights for years. Now, under the provisions of the Submerged Lands Act, they had to relinquish half of their lease areas. Which half they chose to keep was a matter for their own choice. The Queensland government could be expected to seek to reissue permits for these chosen portions as soon as possible; and if the Commonwealth attempted to upset those rights, it would be difficult for it to do so, no matter what the High Court judgement might be. No more was heard of this, for the time being.

In mid-1975, the Whitlam Government brought in its promised legislation, enabling the declaration of a marine national park over the Great Barrier Reef, and the setting up of an Authority, advisory to the Minister, to deal with the Park. A Great Barrier Reef Consultative Committee would advise the Authority itself in matters relating to the Park, and the Minister, in matters relating to the operation of the Act.

The Queensland Premier was angered by this pre-emption of the High Court judgement. As for us, though we were pleased that the legislation now at least existed, the Act was unsatisfactory in other ways. We had supported the GBR Committee's proposals that the Authority should have real administrative power; an advisory Authority which could be over-ridden by the Minister of the day was another matter. As for the boundaries of the Park, Whitlam himself had announced his intention of making the whole of the Reef a marine park. But there was nothing in the Act to suggest this.

During the rest of 1975, with the High Court decision still not available nothing was done to implement the Act. Meanwhile, like Mr Gorton, the Prime Minister and his government grew more and more unpopular, and the centralism that had defeated Mr Gorton was one of the factors in this.

In November, the whole future of the Reef was once again plunged into darkness and question. The Governor-General's action in dismissing the Whitlam Government, and the landslide election that followed, put the future of the Act into the hands of a government pledged to return as much power as possible to the states, and one whose sympathy to conservation questions was soon demonstrated to be very small indeed. The Heritage Commission set up after the acceptance in principle of the report of the Inquiry into the National Estate was hamstrung early in 1976 by the removal of its staff, the Department of Environment was merged into a wider department, and the future for all conservation matters looked, at best, bleak.

It was not until after the election – in which he himself had played a fairly considerable role, by his advice to the Governor-General – that Sir Garfield Barwick released the long-awaited High Court judgement. As expected, this went in favour of the Commonwealth's jurisdiction over all waters offshore below low-water mark.

Yet in March 1976, the Fraser Government found itself still faced with serious problems in the matter of offshore waters. This time they were international rather than national. The series of conferences on the Law of the Sea and offshore ownership was reaching no finality, bogged down in just such squabbles between nations as Peter Scott had predicted to me eight years before. In its claim to the 200-mile limit for exploitation of offshore waters, the Great Barrier Reef's future was being emphasised by Australia as a bargaining point.

But what exactly the Fraser Government intended doing with its authority was still unclear. The oil companies had been held at bay since

1970, first by the appointment of the Royal Commission, then by the long delay in the issuing of the report. The trades unions maintained their black – or green – ban on drilling. But little offshore exploration was being done anywhere in Australian waters, and the clamour for more oil to be found was rising.

Every conservation body in Australia, and many people not members of any, were ready to denounce any damaging moves on the Barrier Reef. No government, not even that of Queensland, was ready (at least near election time) to foreshadow oil drilling in Great Barrier Reef waters. On the surface, at least, it looked as though the long battle might have been effectual. Yet nobody was congratulating anybody. The future was too uncertain, the commitments too vague.

13

Four Years of Looting

Yet it is not only the questions of oil-drilling, mineral recovery from the Reef's waters, or limestone mining, that threaten those waters and the complex of magnificent reefs that they wash and support. The battle against oil has been paralleled by battles against other dangers, and their outcome is as important for the Reef's survival.

The chief culprit in the deterioration of offshore marine ecosystems, apart from oil, is the increasing and myriad forms of pollution that pour into streams, rivers, estuaries and coastal seas. Farms and cities, towns and industry – all are polluters; tourist islands and offshore shipping all add their quota. And in Australia, including Queensland – indeed, especially Queensland – there is very little control over such pollution.

What came of the Premier's promises, after the Albert by-election in 1970, to set up a controlling body to monitor and plan Queensland's environment?

In September 1970 an Environment Control Council was announced. The mouse had emerged from the rhubarb patch. The Council was to be advisory only, and to be made up of representatives of already-existing departments chaired by the Co-ordinator-General. It was to be concerned mainly with questions of 'development'; it could meet as often, or as infrequently, as it liked; it had no provision for independent membership or for public objection to proposals that affected the environment. No biologist member was provided for; conservation bodies were not to be represented.

Introducing the legislation, the Premier claimed that Queensland was 'leading the nation' in the fight against pollution and to 'retain the natural beauties' of the state. He was determined, he said, that pollution would not become a serious problem in Queensland. We had too much reason to know that it was already so.

In the debate, the Opposition had time to make a few points, though the legislation was rushed through. Forty million gallons of sewage, they said, was being poured every month into offshore waters from Cairns alone. Run-off sewage, pesticides and agricultural chemicals from hinterland farms and the Atherton Tableland made the Barron River one of the most heavily polluted in northern Queensland. Fish had already disappeared from the river. The Fitzroy River was carrying a million and a half gallons of primary sewage from Rockhampton every day. The Main Roads Department was using local creeks to wash out its roadside-spraying poison tanks, the Railways Department was pouring oil, dieselene and detergent into coastal rivers, and both departments were to be represented on the Environment Control Council. There was no control over mining and waste disposal, over the use of agricultural chemicals and poison sprays, or over soil erosion.

The Environment Control Council's effect on the pollution pouring into offshore waters was to be practically nil.

In 1971 a Clean Waters Act was passed, with provisions that were so lenient and so long-term in taking effect that it is still largely inoperative.

Said the Public Interest Research Group (a Nader-inspired body based at the University of Queensland), in its Report in 1972,

> What becomes quite apparent … is that the official bodies (whether they be legislative or administrative) believe that, fundamentally, all is well in Queensland … These sorts of beliefs are quite patently reflected in the piecemeal legislation that characterises our area of study, and by a concerted unwillingness to attach any real importance to the claims of consumer and conservation bodies.

There are certain 'special agreements' between the Queensland government and some industrial enterprises which exempt these enterprises from complying even with the mild demands of the Clean Waters Act. Of these large and powerful enterprises (mostly overseas-owned), one was to have special influence in delaying and altering the plan for the new Marine Science Institute at Townsville. This is Freeport Queensland Nickel Inc., whose nickel smelter at Yabulu, just north of Townsville, was the subject of such special legislation. The Greenvale

Agreement Act of 1970 gave the companies concerned what amounted to a licence to pour four million gallons of effluent daily into Halifax Bay near Cape Pallarenda. This was the chosen site for the Institute.

Sulphuric acid, nickel salts and ammonia – all highly effective marine pollutants – are piped out into the Bay in this effluent. How far these toxic pollutants would be carried offshore, and around the shores of Magnetic Island, was problematic. Toxic solid wastes are dumped over 5,000 acres of land near Yabulu. This is a high-rainfall area, with much flood run-off. The Act made no provisions against air pollution from the smelter.

Once these facts were known, it was clear that to establish the Institute at Cape Pallarenda would be foolish. Professor Burdon-Jones, the Institute's chief scientific advocate, had to give up his ideas for the site. As things stood, he pointed out, the Institute might have to import clean sea-water from elsewhere for its aquaria and experiments.

The building of the Institute was politically linked with the name of Mr Gorton, now no longer Prime Minister, and with the legislation claiming Commonwealth authority over territorial waters. Therefore, on his fall, and with the new Prime Minister, Mr McMahon, anxious to propitiate the Queensland government, the establishment of the Institute fell into abeyance. The new site chosen, Cape Cleveland, was also subject to pollution, particularly if settlement extended there from Townsville. A complicated battle ensued to have the Cape Cleveland area declared a national park in order to prevent this. In the event, the Institute is still little more than a plan and a few research staff and projects. The Whitlam government, which had also interested itself in the proposed Institute, was unable to get much further with the actual establishment of the Institute because of these delays.

It seems unlikely that, even with an onshore national park to buffer its area, the scientists will finally have much clean sea to operate in. In an article, by Dr Walt Westman, 'Queensland's Biggest Environmental Scandal: The Greenvale Nickel Refinery', the story of the scientists' attempts to convince the industry is told. 'Never before,' he writes, 'have so many scientists produced so much evidence on the environmental effects of a Queensland project, only to have their warnings ignored. The Greenvale nickel project was the subject of several scientific studies by company-hired consulting firms, including a year-long study done by a Sydney firm on potential water pollution effects. In addition, fourteen independent scientists from James Cook University and the University of Queensland ... discussed the operation of the plant with the company, and made further calculations of environmental risk from the data obtained. The results were remarkably clear-cut considering the usually diffuse nature of environmental problems.'

Apart from air pollution, which would certainly be high, the effluent to be discharged into Halifax Bay would clearly threaten fisheries in the area. The permit to issue this effluent was approved by the Queensland Water Quality Council in spite of what scientists regarded as utterly inadequate data, deficiencies in the calculations, and lack of testing on the effects on marine fauna.

Perhaps it would have been better for Townsville to keep its fisheries and to have a major Marine Science Institute operating in clean waters.

But the nickel smelter is far from being the only factor in pollution of offshore waters in the Barrier Reef region. Most of the North Queensland coastline, especially where the Barrier Reef most closely approaches it, is a narrow shelf of fertile high rainfall soils, backed by high mountains and intersected by short and rapid streams and rivers. High levels of fertiliser and pesticide use, erosion of soils, effluent from sugar mills and towns, and mining wastes, as well as from industrial concentrations such as those at Gladstone, all play a part. Near-shore reefs suffer accordingly, and pollution will increase. Though so far, the scientists of the Marine Institute have found the offshore waters fairly clean, changes in water quality may finally affect parts of the Reef even far from the source of the trouble.

Florida's coral reefs are reported to be dying. Inshore waters off south Florida have lost their clarity. A writer in the journal of the American Museum of Natural History describes them as now 'opaque, yellowish green or brown from eutrophication of suspended non-living material'. He gives a dismal list of the possible sources of this pollution, including not only huge sewage loads from increasing settlement, dredge-and-fill operations by developers, silting and turbidity, but increased loads of such chemicals as PCBs, pesticides, detergents and heavy metals.

Many of these factors already apply in Queensland. Even if the Reef is given marine national park status and mining and drilling are kept out, the Reef will still be under threat from the coastal settlements and industries, and from tourism and tourist islands. No Authority set up to protect the Reef can do so, unless onshore pollution can also be controlled.

The onshore mangrove areas of Queensland, the most extensive left on Australia's eastern coast, are under threat too. These are the breeding areas which produce most of the commercially important fish, shellfish and crustaceans along the coast, and are also the main source of many other species. The relationship of these areas to the Reef, in producing species which may migrate to deeper waters, is still little known.

Mangroves consolidate coasts and estuaries and hold them against erosion. Unfortunately, they are also mosquito-ridden, and reviled by all Shire Councils with an interest in tourism and settlement. Dredge and infill operations for development are therefore usually very welcome to local authorities, and they result in silting which can spread far from the dredged areas. Moreover, mangroves are especially vulnerable to oil pollution, and so are the eggs and larvae of the organisms that breed there.

Queensland's mangrove communities are extensive and rich, even so. The protection given to the coast by the Barrier Reef itself has allowed the build-up of such areas, particularly in sheltered bays and estuaries. Marine biologists concerned with the possibilities of fish and oyster-farming consider that this kind of sea-food production could be made a very important source of income to the state, but at present there is very little sea-farming. The state's chief inspector of fisheries, in evidence to the Royal Commission, said that oyster-farming could become important; oysters grow faster in warmer waters, and there are possibilities that the northern blacklip oyster could become a significant commercial species beyond the range of southern species.

Pollution, whether by oil or chemicals, would make this a precarious enterprise. The onshore drift of currents in the Reef area would mean that any offshore oil pollution would tend to land in the mangrove areas. Oyster spatfall could be affected seriously. In oil industry areas overseas, such as Louisiana, oyster farmers have to resort to using seed oysters or artificial cultivation – a much more expensive method than natural spatfall.

In fact, the pollution of mangrove areas is as important to Queensland as the pollution of the Reef itself. Currents will not wash away oil in the slow and shallow waters of mangrove areas, as they do in deeper waters; eggs and larvae of commercial species, like the famous Queensland mud-crabs and many fish species, cannot escape by swimming away, as adults may do.

One potentially highly productive part of the North Queensland coast is Princess Charlotte Bay – a wide, shallow, well-sheltered bay or sound so far north that at present it has almost no settlement and is therefore unpolluted. Ironically, the area is almost wholly under oil-drilling permit, held by the company in which the Premier holds a high share interest.

As for the controversy over the Crown of Thorns, it still goes on.

While the Whitlam Government was in power, a good deal of research into the problem was set going. Much of it is now being cut down. Independent observers, such as Dr Endean and teams he has taken to the Reef, say that the southward movement of the starfish goes on and that it is only a matter of time before most unaffected reefs are reached. The argument that the increase in the starfish is a cyclic phenomenon has had some support from the finding of remnants of the starfish in cores taken from Reef sediments. But whether this increase is greater than any in the past is still unknown.

Certainly, what the Reef suffers from now is mostly human influence. Apart from pollution, the final blame, if the Reef shares the fate of Florida's corals, lies with us. Tourists, even when they know that corals and shells are protected by law, still take them; they turn boulders and kill eggs and larvae of many species by leaving them exposed to the sun, they drop litter, go spear-fishing and fish out whole reefs of the desirable species – we all know what people are like, for we have done such things ourselves.

But the governments that should be protecting the Reef are not doing so. The final responsibility belongs to them.

The Reef's fate is a microcosm of the fate of the planet. The battle to save it is itself a microcosm of the new battle within ourselves. So this is not just a story of one campaign. The human attitudes, the social and industrial forces, and the people who in one way or other took their part in the campaign, represent a much wider field, and one in which the future of the human race may finally be decided.

14

Finale Without an Ending

This unadorned, bare chronological account of an engagement that lasted through years does not give much hint of the causes and motives of the devotion, some might say obsession, that drove a few people to resist at any odds the commercialising of the Great Barrier Reef. Rather than dramatising our encounters, I have chosen to give the facts and little more.

Some of us who worked in Brisbane had not even seen the Reef. John Büsst, of course, knew it well. I myself had seen only a very small part of it, in the fringing reef of Lady Elliott Island many years before the battle started. But when I thought of the Reef, it was symbolised for me in one image that still stays in my mind. On a still blue summer day, with the ultramarine sea scarcely splashing the edge of the fringing reef, I was bending over a single small pool among the corals. Above it, dozens of small clams spread their velvety lips, patterned in blues and fawns, violets, reds and chocolate browns, not one of them like another. In it, sea-anemones drifted long white tentacles above the clean sand, and peacock-blue fish, only inches long, darted in and out of coral branches of all shapes and colours. One blue sea-star lay on the sand floor. The water was so clear that every detail of the pool's crannies and their inhabitants was vivid, and every movement could be seen through its translucence. In the centre of the pool, as if on a stage, swayed a dancing creature of crimson and yellow, rippling all over like a wind-blown shawl.

That was the Spanish Dancer, known to scientists as one of the nudibranchs, a shell-less mollusc. But for me it became an inner image of the spirit of the Reef itself.

As the battle for the Reef progressed, all of us who were fighting to keep those crystal waters from sacrilege became welded in a very deep companionship, and that in itself helped to keep us at work. But perhaps all of us had some such image to hold and to inspire us when we thought of the shadow that menaced the Reef.

Since the battle began, two of us have died. Arthur Fenton, our always too-willing secretary, and one of the most hard-working of us all, never fully recovered from the overstrain of those years. Taff had been a quiet mainstay, one of those who never ask for honour.

His death, not very long after he had had to retire from the secretary-ship, darkened and saddened the battle; much of what we had managed to achieve was due to him.

As for the man whose energy and devotion had first sparked off, and largely continued, the fight itself, he too is no longer here to be honoured for it. I had been distressed at the change in John, when I last saw him in Brisbane during the preparations for the Royal Commission hearings. He never gave his evidence to the Commission which his work had brought on. Though he was clearly a sick man, we were all deeply shocked and moved when early in 1971 he quietly died at Bingil Bay.

Barry Wain, the young journalist who had covered the Ellison Reef case four years before, had learned to admire John's forceful fearlessness, his wit and his deep appreciation of the beauty of the north where he had lived so long. In an obituary article, Wain wrote:[7]

> With the death of John Büsst of Bingil Bay, the Australian conservation movement lost an able, dedicated strategist and the human race one of its finer members. He was a man who believed passionately in conservation and was truly outraged at the thought of Nature's destruction.
>
> John will be remembered most for his efforts to protect the Great Barrier Reef ... His interest in keeping the Philistines at bay was much broader than that. The other interest simply failed to attract the same publicity. To want to save the rain forests of northern Australia is to be dismissed as a crank in Queensland. Especially if you are a rat-race drop-out, existing hermit-like and content in the 'wilds' of the north ...

[7] *Nation*, 1 May 1971.

Influential supporters were hard to find in late 1967 when an individual sought a lease to mine eighty-four acres of what he described as dead coral, on Ellison Reef... Where were the politicians then with their indignant outpourings, their dramatic confrontations, their vote-catching headline-hunting stands against environmental outrage? Just the sounds of silence as a local Stipendiary Magistrate, doubling up as a Mining Warden, passed judgment in this important test case in a drab country courthouse. The full story of the successful opposition case, organised almost single-handed by John Büsst, remains to be told.

So, of course, did the rest of the story, and of his leading part in that too – a part which the newspapers had mainly missed. Years before I had told John, a propos of one of his quiet triumphs, when the Great Barrier Reef Committee decided to abandon its stand for the 'controlled exploitation' of the Reef and come out to join the conservation groups in opposing oil-drilling, that one day I would write that book.

'Compromise was not part of his approach,' wrote Barry Wain:

the phrase he detested most was 'controlled exploitation', because they were the words of compromise. 'This I regard not as a scientific contribution to the argument but as a half-hearted sop to the exploiters,' he wrote, 'a not very courageous attempt to save some meagre portions of the whole reef from exploitation. I will have none of it ... The control of exploitation, once it has begun, exists only as a myth in the minds of those who advocate it.'

The Reef was above politics. Interviewing both the Prime Minister, then Mr Gorton, and the Opposition Leader, Mr Whitlam, on his home ground, John stressed it was 'a matter of international concern, too important to be made a political football or subject to parochial State interests.' He wanted to see it survive for succeeding generations, not just of Australians, but for all mankind.

Wain's article was headed 'The Bingil Bay Bastard'. John had awarded himself that title after one of his less-known successes. The Army had tried to take over a tract of the splendid rainforest near its tropical trials establishment, for reasons unnamed. John had swung into battle in his inimitably polite, even amusing, but dogged way. It took all his wiles and persistence to discover that it was wanted for defoliation experiments in connection with the Vietnam War. But, as Barry Wain closed his article, 'the Army lost that one to the Bastard'.

Have the oil companies lost the Reef to his valiant opposition, too? That remains for Australians to say. But whatever may happen to the Great Barrier Reef, the battles he waged for it will remain to the credit, not just of John himself, but of all those who try as best they can to save something beautiful and alive from the wreckage of industrialisation and the profit motive.

He has a stone on a hillside in North Queensland, between the great view of the seas of the Barrier Reef in front of his old home, and the piece of rainforest he gave in his will to the James Cook University of Townsville. At the request of his wife, Alison, and the Innisfail branch of the Wildlife Preservation Society of Queensland, I wrote a few words to be engraved on it, and I am proud that they are there.

JOHN BÜSST
ARTIST AND LOVER OF BEAUTY
WHO FOUGHT THAT MAN AND NATURE MIGHT SURVIVE

Postscript

In July 1976, a first step was taken towards implementing the new Whitlam legislation on the Great Barrier Reef, when six members were appointed to the Authority set up under the Act, to administer the Commonwealth's responsibilities over the area, including that of setting up a marine park. What will be its boundaries – will it include the areas designated by the majority view in the Report of the Royal Commissions as 'recommendations for permitted drilling areas', or only those in which the Commissions' majority view recommended that drilling should not be permitted?

If the first, then we can be confident that in effect the Whitlam government's stated intention of excluding oil-drilling from the whole Reef area is to be honoured. If the second, then the marine park area will include at least:[8]

1   12°S. to 18°S. within the outer line of reefs: approximately from Cape Grenville to Dunk Island. This region extends over 'more than 500 miles of coastline where the outer line of reefs is at its closest to the land and the Channel studded with reefal features. The area includes, especially at its southern end, many easily accessible islands of outstanding beauty'.

2   18°S. to 19°S. west of 147°E. (This area includes Hinchinbrook Island and Palm Island.)

[8] Report of the Royal Commissions into exploratory and production drilling for petroleum in the area of the Great Barrier Reef, para. 3, p. 7.

3 19°S. to 20°S. west of 148°E. (This is within the general sea area of the Institute of Marine Science at Townsville.)

4 20°S. to 21°S. west of 150°E. (This includes the Whitsunday, Cumberland and Northumberland groups and the Swain Reefs – the latter reefs contain the area where drilling took place, without success, before the campaign against oil-drilling began.)

5 21°S. to 22°S. west of 150° 31′E. (This includes the Capricorn and Bunker Groups, where there are already tourist centres such as Heron Island and cays which can easily be reached and are being used by scientists for marine research.)

6 23°S. to 24°S. west of 152° 50′E.

How likely is it that oil might be found in the Great Barrier Reef area? Estimates given to the Royal Commissions varied, but presumably the exploration permits which were granted originally by the Queensland government delineated pretty well all the possibilities. There are seven sedimentary basins within the Great Barrier Reef area, and if there is oil anywhere in the Reef area, these are the likeliest places for it to be found. From north to south, these basins are:

The Papuan basin (north and east of Cape York). This was rated by the Queensland Geological Survey Branch of the Mines Department, and by the Geological Branch of the Bureau of Mineral Resources, Geology and Geophysics in Canberra, as a 'good' prospect. But it also contains the Torres Strait Islands – and at present the disputed question of the boundary between Australia and New Guinea remains unsettled (September 1976). The strongly held view of the Islanders, who have already been subjected to the effects of the *Oceanic Grandeur* tanker spill, is against any oil or mineral exploitation in their fishing waters. The view of oil companies, and probably of both the state and the present Federal governments, is clearly otherwise. Some offshore exploration is going on already within the undisputed waters of the Papuan Gulf which certainly belong to Niugini.

The Laura basin, offshore from eastern Cape York south of *The Papuan basin.* This is rated as a 'fair' prospect. It contains the Princess Charlotte Bay area, in which Oilmin, the company part-owned by the Premier of Queensland, holds a very strong interest. Princess Charlotte Bay, though still isolated, has a high potential for the development of a marine farming industry; indeed, its very isolation and consequent lack of pollution would make it very suitable for the development of fish-farming and an oyster industry might be based on the northern black-lip oyster.

192

The Halifax basin, north of Townsville, was not given a rating, and its potential was said by the Royal Commissions to be unknown. *The Proserpine* or *Hillsborough basin* lies offshore between Bowen and Mackay. Its potential was rated as 'poor to fair'.

The Styx basin lies off Broadsound, and is rated as 'poor'.

The Capricorn basin lies eastward of Heron Island and is rated as a 'fair' prospect. But its tourist importance, and the accessibility of its cays and reefs, would make any major or continuing pollution here very visible indeed. The establishment of the Heron Island marine research station, and the amount of research already being carried on, is another factor.

The Maryborough basin lies east of Bundaberg, and was rated as a 'fair' prospect. Only the southern part of the basin attracted any application from oil companies.

All in all, then, the prospects for finding oil south of the Papuan basin were not rated highly. How should we estimate the chances of great pressure from oil companies to exploit the Reef's possible oil resources?

Firstly we have to consider the stated determination of the Queensland government that the Reef shall be drilled. (The Mines Minister has said in the past, without reservation, that the Reef will be drilled.) The High Court decision in December 1975 in effect handed jurisdiction in the matter to the Federal Government. The present government has not made any firm statement of its views on oil-drilling in the Reef province. But its bias will certainly be towards the oil companies wherever possible.

Late in August 1976, the Federal Attorney-General, Mr Ralph Ellicott, attended a meeting of the standing committee of Attorneys-General. He told the committee that he was 'very grateful' to receive from the Western Australian government a submission for the extension of the states' jurisdiction into offshore regions, and that the Commonwealth 'would be happy to entertain this thought if it were put by each Premier to the Prime Minister'. Sir Charles Court of Western Australia promptly wrote to the Prime Minister asking for further exploration of the question as a matter of urgency.

If, as the Premier asked, sovereignty over the offshore areas were handed back to the states, we would be back to square one, for the Petroleum (Submerged Lands) Act which designates the states the right to administer exploration permits and production licences for petroleum had not been repealed.

The question of whether this could be done through section 123 of the Constitution, which deals with the question of altering state boundaries through mutual agreement between the Federal and state governments, or by the provision for constitutional amendment, was thought by the chairman of the Offshore Laws Committee of the Attorneys-General, to be easy. The US example was there, and 'had caused no embarrassment to the United States internationally'.

The shadow minister for minerals and energy has warned that if a Labor Government took office, it would review any offshore permits or leases the state governments might grant. While any Labor government would certainly scrap the Submerged Lands Act, it is very unlikely that any such government in the foreseeable future would have control of the Senate. And by that time, there would probably be a *fait accompli* on the Great Barrier Reef.

The whole Federal-states issue is at the very core of the problems of Australian government. With the two biggest (in area) states straining against the centralism which the Federal Government under Mr Gorton first brought into the limelight, and the Whitlam government took further, questions of offshore sovereignty have always been the hottest issue. The stake for Western Australia is very big, with petroleum and natural gas already established; for Queensland, its friend and rival, the stake is the Great Barrier Reef – and any oil or gas discovered offshore of the north and west of Cape York, or in the Gulf of Carpentaria off Queensland's boundaries.

The stake for the Torres Strait Islanders is their expressed wish to have no oil industry established in the Strait. And the stake for Australia is the question of good government, international recognition in the Law of the Sea Conference, and, again, the Great Barrier Reef.

In August, too, it was announced that the chairman of the Foreign Investment Review Board and the full-time representative of the Treasury on that Board would be making a world-wide trip to attempt to attract more funds for offshore exploration and development.

What will the bargain be? While the estimate of the potential of the Great Barrier Reef region as a likely oil-producer was far from high, the oil companies which originally held permits in the Reef area, and had spent money in survey and other costs did not enjoy either the bad publicity given them over the Reef issue, or the frustration of their plans. Certainly the Queensland government did not. It would be a triumph for both to have the Federal Government back down on the Whitlam legislation and the Whitlam plans to turn the whole Reef area into a marine park in which no oil-drilling would be allowed.

In September 1976, then, it seemed all too possible that the whole of the battleground might revert to the positions held in 1969, and that the six years following the imposition of the trade-union ban on Reef drilling had been no more than a respite for re-grouping by the forces. And meanwhile, the Queensland state government was threatening to legislate to prevent any such bans from being made, or becoming effective ...

In November, however, the Prime Minister was forthright in his rejection of the idea of extending states' jurisdiction offshore. Mr Ellicott had apparently spoken too soon.

But during November, too, the Fox Report on the question of mining and exporting uranium was issued. Its very strong reservations on this, and the rising public disquiet on the whole matter, had many implications for Australia's overseas income, supposing that the Fraser government accepted the need for restrictions on uranium export. That point would certainly be made by overseas oil companies, which were likely to be already pressing for Great Barrier Reef exploration.

Indeed, the now-diminished and threatened Department of Environment, Housing and Community Development was already looking into the question of oil-drilling on the Reef again. Undoubtedly the Queensland government, which still stood to gain considerable royalties from offshore oil-drilling, whatever the sovereignty over Great Barrier Reef waters, was also pressing its case.

The battle had been muted during 1976. Not only the Labor Party had expressed itself against oil-drilling, but before the election the Liberal Party, at least, had said that it was not in favour of exploiting possible oil reserves in Reef waters.

But the report of the Royal Commission, with its majority recommendation that drilling could be undertaken in some parts of the Reef at least, could provide the government with at least an excuse for reversing that rather ambiguous stand.

In January 1977, it became clear that the states were far from accepting the Prime Minister's views on offshore jurisdiction and the judgment of the High Court. The Premier of Western Australia opened for the states with an attack on Mr Fraser's 'centralism' in refusing to hand back at least some control over the territorial seas and seabed. One of his complaints was that the Federal government had decided against inserting a clause in a convention on national parks and wildlife, which would have reserved to the states the power of declaration of marine national parks.

Since this question seemed rather irrelevant to the interests of Western Australia, where no major marine national park appeared to be even being considered, it was easy to conclude that this particular grievance was being voiced on behalf of another recalcitrant state, and that that state was Queensland.

Once again the fate of the Reef lay in the hands of a Prime Minister, and this time, one otherwise committed to the view that states should control their own affairs, and vulnerable to pressure from oil lobbies and from his own party.

1977, then, might prove to be the decisive year for the Reef's whole future. Those coral gardens, two years after the issuing of the Report of the Royal Commission, were still a battleground.

References, Sources and Bibliography

Newspapers
Canberra Times.
Courier-Mail, Brisbane.
Daily Telegraph, Brisbane.
Financial Review.
Nation Review.
Santa Barbara News Press.
Sunday Telegraph, Sydney.
Sunday Truth.
Sydney Morning Herald.
The Age, Melbourne.
The Australian.

Journals
American Oil and Gas Journal, Tulsa, September 1973.
American Society of Mammalogists, Journal, November 1968.
Eco-Info, Journal of the Queensland Conservation Council, November–December 1974.
Marine Pollution Bulletin, vol. 2, no. 4, 1971.
Nation, 1 May 1971.
Operculum, a publication of the Queensland Littoral Society and the Australian Littoral Society (NSW Division). Various.
Search, Australian & New Zealand Association for the Advancement of Science. Various.

Living Earth, vol. 14, no. 2.

Trend, the Labor News Magazine. February 1970.

Wildlife in Australia, quarterly magazine published by the Wildlife Preservation Society of Queensland. Various.

Newsletters, Minutes, Annual Reports

Australian Conservation Foundation. Newsletters, annual reports. Co-ordinating Committee of Conservation Bodies appearing before the Royal Commissions into Petroleum Drilling in Great Barrier Reef Waters. Minutes. Littoral Society of Queensland. Newsletters and special bulletins. Save the Reef Committee, Brisbane. Minutes and annual reports. Wildlife Preservation Society of Queensland. Minutes, annual reports, newsletters, special bulletins.

Special Reports, Submissions, Evidence

Australia. Committee of Inquiry into the National Estate, *Report on the National Estate*. Australian Government Publishing Service, Canberra, 1974. And submissions thereto by the Great Barrier Reef Committee (Doc. 147) and Save the Reef Committee (Doc. 203).

Australia. House of Representatives, *Debates*.

Australia. House of Representatives, Select Committee on Wildlife Conservation. Transcript of evidence taken at Brisbane Tuesday 15 February 1972.

Australia. Senate, Select Committee on Off-Shore Petroleum Resources, various submissions.

Australia. Senate, *Water Pollution in Australia*. *Report of the Select Committee on Water Pollution*. Also various submissions.

Australian Academy of Science, *Acanthaster planci (Crown of Thorns Starfish) and The Great Barrier Reef*. Report no. 11.

Australian Conservation Foundation. *Environmental Pollution*. Special publication no. 6.

————, *The Future of the Great Barrier Reef*. Special publication no. 3.

Innisfail. Warden's Court, Transcripts of evidence given at the hearing of an application for the mining of limestone on Ellison Reef, 1967.

Public Interest Research Group, *The State of Queensland*. A Report on Pollution.

Preliminary Report on Conservation and Controlled Exploitation of the Great Barrier Reef (Chairman: H. S. Ladd), Government Printer, Brisbane, 1968.

Report on the Grounding of the Oil Tanker Oceanic Grandeur *in the Torres Strait on 3 March 1970 and the Subsequent Removal of Oil from the Waters.* A. 32-1970.
Royal Commissions on Petroleum Drilling in Great Barrier Reef Waters, summary and conclusions, various submissions and statements.

Books

Bennett, Isobel, *The Great Barrier Reef.* Lansdowne, Melbourne, 1971.

Brown, Theo, & Wiley, Keith, *Crown of Thorns.* Angus & Robertson, Sydney, 1972.

Claire, Patricia, *The Struggle for the Great Barrier Reef.* Collins, London, 1971.

Connell, D.W., *Water Pollution in Australia.* University of Queensland Press, St Lucia, 1974.

Legalized Pollution: The P.I.R.G. Report on Pollution Control Laws in Queensland. University of Queensland Press, St Lucia, 1973.

Martin, Angus, *Pollution and Conservation in Australia,* Lansdowne, Melbourne, 1971.

Marx, Wesley, *The Frail Ocean.* Ballantine/Sierra Club, 1967.

Serventy, V., *The Great Barrier Reef.* Golden Press, Sydney, 1972.

Webb, L. J. *Environmental Boomerang.* Jacaranda, Brisbane, 1973.

Other Sources

Wildlife Preservation Society of Queensland, correspondence files.
Author's correspondence.
Dr R. H. Endean, Department of Zoology, University of Queensland. Personal communication.

Index

Holt, Harold, 4, 18, 27, 114, 162
Holt, Dame Zara, 27
Housewives' Reef Preservation
    Committee, 115

Ingham, 32
Innisfail, 4, 6, 9, 10–12, 32, 65, 79, 83, 86,
    95, 99, 130, 190
Institute of Marine Science, *see* Townsville
International Union for the Conservation
    of Nature, 31, 83, 136

Japex Ltd, 86, 89, 90, 97, 105–7, 109–12,
    115, 121–2, 127–9, 132. *See also* Ampol-
    Japex.
Jacobs, Professor Wilbur, 73
James, Peter, 13n

Keeffe, Senator, 105
Kodak Ltd, 11

La Macchia case, 78–9, 118, 151
Labor Party: Federal, 52, 54, 77–8, 88, 97,
    114, 115–16, 118, 119, 155, 158, 172,
    194, 195; State, 53, 119, 129, 131–2,
    148, 165–6, 176. *See also* Whitlam
    Government.
Ladd, Dr H. S., 25, 29, 30, 31, 34, 35–8
Lady Elliott Island, 1, 187
Laura basin, 192
Leslie J. Thompson, 140, 143–4, 148
Liberal Party: State, 74–5, 76–7, 83–4, 115,
    129–30, 131–2, 166, 176; Federal, 118,
    145, 162, 172, 195
Limestone mining, 6–14, 18–19, 24, 34,
    44–5, 97
Lippiatt and Co., 151, 154, 161
Littoral Society of Queensland, 5–6, 7, 9,
    11–13, 23, 24, 31, 36, 50, 52, 61, 62, 65,
    72, 73, 80, 83, 84, 87, 88, 96, 111, 125,
    134, 135, 152, 168

McEwen, Sir John, 129, 130
Mackay, 53, 55, 58, 59, 69, 81, 89, 98,
    106–7, 120, 152, 170, 193
McLennan, Freda, 168
McMahon, William, 183
McMichael, Dr Don, 10–12, 40, 41, 144
Magnetic Island, 101, 166, 183
Mangrove areas, 21, 184–5
Marine national parks, 2–3, 6, 7, 16–17, 27,
    33, 68, 130–1, 175, 177–9, 184, 191
Maryborough, 47; basin, 193
Mather, Dr Patricia, 23, 26, 76, 90, 137,
    146, 152, 168

Maxwell, Professor, 62–3, 66, 146
Mines Minister, *see* Camm, R. E.
Moffitt, Ian, 81
Monkman, Noel and Kitty, 2
Morley, I. W., 48, 57
Moroney, V. J., 153

Nation, 188
National Party, *see* Country Party
Navigator, 97, 98, 99, 100, 102, 103, 105,
    106, 110, 112, 121, 128, 132, 143, 170
Nixon, President, 104, 108
Northumberland Group, 192

Occidental of Australia, 154
Oceanic Grandeur, 140–1, 144, 148, 173, 192
Offshore jurisdiction, 15–19, 25, 38,
    49; conflict between States and
    Commonwealth, 63–4, 76–9, 82–3, 92,
    98, 112–19, 136, 142–5, 162–5, 172–3,
    176–7, 178–9; High Court judgement,
    179, 193
Offshore legislation: Commonwealth,
    16–19, 43–4, 82, 141–2, 143, 144, 162–5,
    172–3, 176–7, 178–9, 193–4; State,
    16–17, 28, 59
Oil drilling, permits for, 8, 24, 27, 32–3, 47,
    52–3, 58–9, 71 (map), 95, 106, 113–14,
    115–16; effects of, 21–3, 66–73 *passim*,
    86–7, 89, 95–6; public inquiry (later
    Royal Commission) into, 110–19 *passim*,
    123–7, 134–9, 142, 149–58, 160–1,
    166–72, 174–5, 177–8, 191, 192, 195; in
    Repulse Bay, *see* Ampol
Oil spills, 32, 45, 47–8, 51, 52, 53, 57, 64,
    70, 75–6, 81, 133, 140–8 *passim*, 166, 173
Oilmin, 192
Orme, Dr, 54
Outer Barrier, 37

Pacific-American Oil and Shell
    Development, 33
Palm Island, 72, 191
Papuan basin, 192, 193
Patterson, Dr Rex, 33, 52, 54, 115
Piesse, Dick, 45, 61
Planet Exploration Company, 154
Planet Metals, 28, 144
Pollution, 6, 45, 72–3, 96, 104, 132–3,
    181–5. *See also* Oil spills.
Princess Charlotte Bay, 47, 69, 106, 164,
    185, 192
Proserpine basin, 193
Public Interest Research Group, 182
Public opinion polls, 83–5, 86, 88–9, 97, 98

Acknowledgements

I have to thank Dr Fred Grassle of Woods Hole, Mass., U.S.A., Mr Vincent Serventy, and Professor Frank Talbot of Macquarie University for permission to quote various passages from their writings, and Dr Geoff Mosley, Director of the Australian Conservation Foundation, for permission to quote from ACF Newsletters. I have not been able to locate Mr Barry Wain for permission to quote from his article in *Nation*, but I think this would have been freely given.

I also acknowledge various brief quotations from editorials and news items in the press, particularly the *Courier-Mail* and *The Australian*.

The map on page 71 is by courtesy of the *Courier-Mail*, Brisbane. The extract from Kenneth Slessor's poem, 'Five Visions of Captain Cook', is reproduced by kind permission of the estate of the late Kenneth Slessor and Angus & Robertson, Sydney, from *Poems*, 1944.